INTERNATIONAL ENVIRONMENTAL DISPUTES

A Reference Handbook

Other Titles in ABC-CLIO's
**CONTEMPORARY
WORLD ISSUES**
Series

Books in the Contemporary World Issues series address vital issues in today's society such as genetic engineering, pollution, and biodiversity. Written by professional writers, scholars, and nonacademic experts, these books are author- itative, clearly written, up-to-date, and objective. They provide a good starting point for research by high school and college students, scholars, and general readers as well as by legislators, business people, activists, and others.

Each book, carefully organized and easy to use, contains an overview of the subject, a detailed chronology, biographical sketches, facts and data and/or documents and other primary- source material, a directory of organizations and agencies, annotated lists of print and nonprint resources, and an index.

Readers of books in the Contemporary World Issues series will find the information they need in order to have a better understanding of the social, political, environmental, and economic issues facing the world today.

INTERNATIONAL ENVIRONMENTAL DISPUTES

A Reference Handbook

Aaron Schwabach

CONTEMPORARY WORLD ISSUES

A B C  C L I O

Santa Barbara, California
Denver, Colorado
Oxford, England

344.04
Sch

Library of Congress Cataloging-in-Publication Data
Schwabach, Aaron.
    International environmental disputes : a reference handbook / Aaron Schwabach.
        p.   cm. — (Contemporary world issues)
    Includes bibliographical references and index.
    ISBN 1-85109-773-2 (hardback : alk. paper) — ISBN 1-85109-778-3 (ebook) 1. Environmental law, International.   2. Liability for environmental damages.   3. Environmental protection—International cooperation.   I. Title.   II. Series:Contemporary world issues
K3585.S388 2006
344.04'6—dc22

                                                                    2005019814

09   08   07   06   10 9 8 7 6 5 4 3 2 1

This book is also available on the World Wide Web as an eBook.
Visit abc-clio.com for details.

Production Team
    Acquisitions Editor: Mim Vasan
    Associate Production Editor: Cisca Louise Schreefel
    Editorial Assistant: Alisha Martinez
    Production Manager: Don Schmidt
    Manufacturing Coordinator: George Smyser

ABC-CLIO, Inc.
130 Cremona Drive, P.O. Box 1911
Santa Barbara, California 93116-1911

This book is printed on acid-free paper ∞.
Manufactured in the United States of America.

For Veronica and Jessica

Contents

Preface and Acknowledgments

The twentieth century, with its dramatic worldwide increases in wealth, life expectancy, and industrialization, saw a dramatic increase in humanity's impact on the environment. Increasingly, people and governments became concerned with environmental problems and realized that many of these problems could only be addressed by international, not merely national, action. Human activity does not need to trigger a worldwide ecological catastrophe in order to raise issues of international law. The opening of an airport, a factory, or a sewage treatment plant in a city on or near an international border may do so. Natural systems are not constrained by national boundaries; if they are to be protected, they must be protected internationally, which requires cooperation and some sacrifice of sovereignty by the countries concerned.

This book serves as a reference for those who want to explore humanity's attempts, up to and throughout the twentieth century and into the twenty-first, to create a workable global regime of environmental protection. It is designed to serve as a starting point for future research; international environmental law is constantly changing and evolving, but the resources provided here will make it possible to locate up-to-the-minute information in a wide variety of areas. For the most part the changes in international environmental law have been for the better, leading to a more coherent and inclusive regime. There have been setbacks along the way, such as the apparent abandonment of the community concept of drainage basin management after World War II or, more recently, the withdrawal of the United States from the Kyoto process. These setbacks may turn out to be positive developments, however; they define the limits beyond which states are unwilling to sacrifice sovereignty and provide a guide to what is actually achievable.

Chapter 1 of this book begins with a historical overview of the development of international law. It divides this development into three historical periods. In the first period, before 1941, international environmental law had no separate existence; although a few environmental treaties existed, environmental protection as a whole was generally incidental rather than a specific goal of international law. That changed with the 1941 decision in the *Trail Smelter* arbitration. Although its importance as a demarcation point may be more obvious now than it was at the time of the decision, the *Trail Smelter* ushered in the modern era of international environmental law, in which environmental protection itself was a goal of international law, and environmental injury became an international wrong for which states could seek redress.

A further landmark event occurred in 1972 with the United Nations Conference on the Human Environment in Stockholm and the creation of the United Nations Environment Programme (UNEP). After 1972 environmental protection became not merely a goal of international law but a legal and administrative field in its own right. Chapter 1 then turns to an examination of this area of law, looking both at the role of UNEP and other international organizations and at the structure and functions of international environmental protection. It examines two controversies that dominate international environmental law today: the North-South controversy and the related controversy over sustainable development. It concludes with a brief look at the principles of environmental ethics underlying the field.

Chapter 2 looks at specific international environmental problems and the degree of success that has been attained in resolving them. In the Gabčíkovo-Nagymaros dispute, for example, two neighboring states followed a highly formal process of dispute resolution through the International Court of Justice. More often, disputes are too unfocused to lend themselves to such formal modes of dispute resolution. At least one of these more fuzzily delineated problems is double-edged: Trade barriers in rich countries help to keep poor countries poor. Some environmentalists think that the environment would be harmed if these countries became rich; others think that, for example, the environment of Lake Victoria would be better off if the people living around it were wealthier. The area around the lake has been deforested, leading to a loss of wildlife habitat, soil erosion, and pollution of the lake. The trees are cut to smoke the flesh of the Nile perch (a fish that is an environmental disaster in its own right) in order to preserve it. Wealthier people would be able to afford refrigerators (and electricity) and would have no need to cut the trees.

The chapter looks at successes such as Antarctica and the ozone treaty regime, as well as failures such as the global climate change regime and ongoing problems such as toxic pollution in the former Iron Curtain countries, ocean and freshwater pollution, and transboundary shipments of toxic wastes. It also looks at an "environment" not usually thought of as such—outer space —and at a dispute over a resource in space.

Chapter 3 looks at problems specific to the United States. The United States has the world's third largest population and largest economy; its actions have major international environmental impacts. The chapter first looks at the role of international law in the U.S. legal system. It then examines attempts to have U.S. laws applied to the actions of U.S. parties outside the United States, at the law governing environmental damage from U.S. military actions, and at a suit by a foreign national in a U.S. court for environmental harm allegedly committed in violation of international law.

Chapter 4 provides a chronology of milestone events in the development of international environmental law, with particular attention to events relating to disputes discussed in detail elsewhere in the book. Chapter 5 provides biographical sketches of a selection of activists and leaders who have helped to shape the development of international environmental law. These are as diverse a group of individuals as any on the planet; the environment, after all, is everyone's concern. They include a New Guinea tribal elder, a Norwegian prime minister, a Japanese TV personality, an internationally famous Nigerian author who was executed for his activism, and several others.

Chapter 6 provides excerpts from a selection of important documents in international environmental law, along with explanatory text. It cannot, of course, provide all of the important documents—but the organizations listed in Chapter 7 and the resources listed in Chapter 8 can. Nearly every primary source of international law is available online, and one of the goals of this book is to enable the reader to find these sources. Chapter 7 describes and provides contact information for intergovernmental organizations, including treaty secretariats, and a sampling of non-governmental organizations. Chapter 8 provides a bibliography, with descriptions of books suggested for further reading; a list of articles, journals, and primary source materials; and a guide to the most comprehensive online resources.

This book would not have been possible without the assistance of a large number of people. I would particularly like to thank my research assistants Krista Schelhaas and Andrea Patten; my editors Mildred Vasan and Cisca Schreefel; Dick Scott, Bill Slomanson, and Ken Vandevelde, for looking over drafts; Thomas Jefferson School of Law, for giving me the time to work on this project; and my family, for their patience in putting up with me while I did. I hope that you enjoy reading this book as much as I did writing it, and I hope that it will serve as a starting point for further research on and discussion of international environmental law.

1

International Environmental Law in Context

The world environment is in crisis. Or is it? One view is that anthropogenic (human-caused) environmental change is now the greatest threat to humanity's continued survival as a species within the global ecosystem. Undeniably, the human race is now consuming many resources at an unsustainable rate; ancient forests and wetlands vanish daily. As a result of human activity, the delicate balance of Earth's natural systems has been upset. Global warming may cause altered weather, including droughts, floods, and an increase in the frequency and severity of hurricanes and tornadoes, as well as inundation of the coastal areas that are home to most of the human race. Species extinctions are occurring at the highest rate in tens of millions of years, with the consequent loss of irreplaceable genetic information. Loss of biodiversity may ultimately threaten the species on which humanity depends for food, either directly or indirectly through increased vulnerability to disease. Humans move more soil, rock, and sand than all natural processes of erosion combined. Rivers, lakes, and even oceans are becoming open sewers; many of the underground aquifers that contain most of the world's available fresh water are being depleted by poorly managed use or contaminated by toxic wastes, sewage, and agricultural chemicals. Overgrazing turns grasslands into deserts.

Yet there is another side to the picture. The quality of the human environment has improved steadily since the industrial revolution. Although famine and disease persist in many parts of the world, for the human race as a whole life expectancy has increased and infant mortality has decreased—dramatically. More people, both in proportion and in total numbers, have access to safe drinking water and uncontaminated food than ever before. The air has gotten cleaner, too, at least in some places where people live; it is hard to look at London today and imagine the dense smog that used to descend on the city for days at a time, killing hundreds of people a day. More landfills are lined and sealed and more sewage is treated than ever before; more countries are banning the use of leaded gasoline, highly toxic pesticides, and other dangerous substances. For the average human being, the environment has never been more conducive to good health.

This rosy view also has its problems, though. One is an aesthetic, nearly metaphysical objection to living entirely within a human-created environment. On any given day billions of people walk only on man-made surfaces, tilled fields, and carefully tended lawns; they spend their days in buildings, in cars, in carefully landscaped and managed parks, malls and plazas. For those who, as Aldo Leopold put it, can live without wild things, this is no hardship. For those who cannot, something beyond price has been lost.

Then there is a more pragmatic concern. Life has existed on Earth without interruption for billions of years. Nothing of which humanity is yet capable can render the Earth unsuitable for *all* life. It is quite possible, however, for human activities to render the Earth unsuitable for *human* life. Humanity is a new species, and has not yet demonstrated any notable evolutionary staying power. We do not know what degree of environmental change might lead to the extinction of the human race, and therefore all major environmental changes—climate change, an increase in ultraviolet radiation reaching the earth's surface, extinctions of edible plants—are cause for concern.

Environmental threats are global, not national. Environmental systems are not restrained by national boundaries. When a polluter in one country dumps chlorofluorocarbons into the atmosphere, the consequent damage to the ozone layer affects the whole world, not just the inhabitants of the polluter's country. When a species becomes extinct, it is lost to the world, not just to the country or countries in which it once lived. For this reason national law alone is insufficient to address the environmental problems facing the Earth; environmental problems are often international in nature, and must be addressed internationally.

Environmental protection has been a concern of international law throughout history; however, prior to the twentieth century it was a minor concern, and customary and conventional international law regarding the environment evolved as a by-product of the development of law in other areas, such as navigation and fisheries.

This chapter begins with a look at the nature and sources of international law. It then traces the history of modern international environmental law from its origins as an incidental element of treaty regimes designed for other purposes, such as regulating access to navigable waterways, to its emergence as a distinct area of international law. The historical portion of the chapter discusses the 1941 decision in the *Trail Smelter* arbitration, generally seen as the starting point of modern international environmental law, and the 1972 Stockholm Declaration on the Human Environment, which signaled universal (or near-universal) acceptance of the principle of state responsibility first enunciated, albeit *in dicta*, in the *Trail Smelter* decision. The Stockholm conference and declaration in turn made possible the rapid development of international environmental law over the last three decades; the history of international environmental law can thus be divided into the periods before, between, and after these watershed events, and the historical portion of the chapter is organized accordingly. The chapter then examines the current state of international environmental law: the role of the United Nations and its specialized agencies; the rule-making, enforcement, and dispute resolution processes common to environmental regimes; and the concept of limited territorial sovereignty. It concludes with a look at the trends and tensions that may shape the future development of international environmental law. These include the tension between the developed nations of the global North and the developing nations of the South, and the related and inevitable tension between environmental protection and development. They also include the environmentalist struggle to have sustainable development and its supporting principles accepted as rules of customary international law; these principles include the polluter pays principle, the precautionary principle, and the principle of intergenerational equity, enunciated in aspirational documents such as the Rio Declaration and embodied in some international agreements. There is also ongoing debate over philosophical approaches to environmental rights, including the idea of environmental rights as human rights.

Sources of International Law

International law is either *conventional* or *customary;* these words are terms of art and do not mean what they might in ordinary English usage. Conventional international law is law set forth and contained in treaties and other international agreements. Customary international law is a set of normative expectations derived from the practice of states as international actors, undertaken out of a sense of legal obligation. In other words, customary international law can be determined by looking at what states do, not because it is necessary or profitable for them to do so, nor from altruism or some other motive, but because they think that international law requires them to do so.

A widely used starting point for determining the sources of international law is the list contained in Article 38(1) of the Statute of the International Court of Justice (ICJ):

- international conventions
- international custom, as evidence of a general practice accepted as law
- the general principles of law recognized by civilized nations
- judicial decisions
- the teachings of the most highly qualified publicists of the various nations

The Statute provides that the last two items (judicial decisions and teachings) are to be considered by the ICJ "as subsidiary means for the determination of rules of law." In any event, judicial decisions and, to the extent that a state actually observes them, general principles of law *are* state practice, and thus can form the basis for normative expectations.

Although some oversimplification is involved, law found in the sources listed in the Statute (international conventions aside) can be grouped together under the heading of customary international law: a set of normative expectations about the behavior of states formed by stated practice undertaken out of a sense of legal obligation. The third item on the ICJ's list, "general principles of law," has traditionally been viewed as a third category of public international law. It can also be seen as a source of "supplemental rules" or a "secondary source of law." This is the approach taken by the Restatement (Third) of the

Foreign Relations Law of the United States. Reference to "general principles" is common in civil law legal systems (those legal systems derived from Roman rather than English law), but it is unfamiliar to lawyers from common law systems such as that of the United States, which may be why U.S. lawyers are reluctant to consider it a separate category of international law on a par with conventions and custom. (Incidentally, the Restatement and other modern treatments replace the Statute's reference to "recognized by civilized nations," a relic of European colonialist thought, with the less judgmental phrase "common to the major legal systems.")

Of the sources discussed in this book, some may be taken as providing definitive statements of international law, others may be taken as providing evidence of customary international law, and still others must be viewed as purely aspirational. These sources include treaties and other international agreements (the sources of conventional international law); decisions of international courts and tribunals; resolutions of the United Nations Security Council and General Assembly; various documents produced by other international organizations; United Nations conference documents; documents produced by nongovernmental organizations (NGOs); and laws, decrees, and judicial decisions of national governments.

Treaties and other international agreements currently in force may be taken as definitive statements of international law. Although treaties to which the United States is a party are part of the law of the United States under Article VI, clause II of the U.S. Constitution, they do not automatically create rights enforceable under U.S. law unless they are self-executing. A non-self-executing treaty creates no rights enforceable under U.S. law unless implementing legislation has been enacted; such a treaty may thus confer an obligation upon the government of the United States with respect to the governments of other countries, but not with respect to its own citizens. Even treaties to which a country is not a party may eventually come to be accepted in the practice of states as stating rules of customary international law, and thus create binding legal obligations even on those countries that are not parties (Bederman, 2001).

The International Court of Justice and its predecessor, the Permanent Court of International Justice (PCIJ), are together known as the World Court, although the term is often used in the press to refer to the ICJ alone. Decisions of these courts in contentious cases are definitive statements of the international rights and obligations of the parties actually before the court with regard to the subject of the dispute. The ICJ may also issue advisory opinions, as did the PCIJ; these have no binding effect but are generally regarded as reasonably reliable indications of the state of customary international law. Decisions of other international courts and tribunals, where jurisdiction is based on the consent of the parties, are also, in ordinary circumstances, definitive statements of the rights and obligations of the parties under international law. Decisions of international courts and tribunals are *not* sources of international law in the sense that a decision of the U.S. Supreme Court is a source of U.S. law; the international courts are not bound to follow precedent (and there is no clear hierarchical ordering among them), but their decisions are generally taken as an indication of international custom (Shahabuddeen, 1996). The lengthy written opinions that accompany the decisions are also invaluable, as they generally include exhaustive explanations of the history, development, and current status of international law relating to the issue presented to the court.

Resolutions of the United Nations Security Council, although not mentioned in Article 38(1), are statements of international law and create binding legal obligations. The Security Council is concerned with the maintenance of international peace and security, however, not with the environment; any environmental effect of Security Council resolutions is incidental.

Documents produced by certain treaty-based organizations, such as the various organs of the European Union or the U.S.-Canada and U.S.-Mexico boundary waters commissions, have legal effect insofar as they are within the limits on the authority granted to those organizations by the underlying treaties.

Documents produced by United Nations conferences or by NGOs have no inherent legal effect; they are at best aspirational statements. Some, however, may find acceptance in the practice of states and ultimately come to be seen as stating obligations under customary international law.

Resolutions of the General Assembly are not in and of themselves statements of international law; their treatment is complex and requires some discussion. Despite some superficial similarities in structure and procedures, the General Assembly is not a legislative body. There is considerable disagreement as to what normative expectations, if any, its resolutions create.

The traditional approach is that General Assembly resolutions have no legal effect; what significance they have as sources of law is limited to the extent to which they express already existing customary international law. At the other extreme, some international legal scholars and political leaders are willing to accept General Assembly resolutions as a source of customary international law. This approach may be particularly appealing to leaders of NGOs and smaller or newer countries whose interests are not well-represented in the existing body of customary international law. In the *Nicaragua* case and, less strongly, in the *Legality of Nuclear Weapons* advisory opinion, even the International Court of Justice has been willing to consider General Assembly resolutions as a source, rather than merely an expression, of customary international law.

An intermediate approach to the treatment of General Assembly resolutions is provided by the decision of the sole arbitrator in the arbitration between the Texaco Overseas Petroleum Company (TOPCO) and the government of Libya. In that case the arbitrator set out four conditions that must be met for a General Assembly resolution to constitute a statement of international law with respect to particular states. First, the resolution must be accepted by all of the groups of states concerned. Second, acceptance must be demonstrated; acceptance requires not only that a state vote in favor of the resolution in the General Assembly but also that the state's practice conform to the resolution. Third, the state must not have objected to the resolution at the time of its adoption; in other words, the recorded comments of the state's General Assembly delegates must be taken into account, as well as the votes. Fourth, a resolution adopted without universal support cannot replace an existing rule of customary international law (Schwebel, 1979; Garibaldi, 1979).

These four requirements may seem to set a very high bar for the acceptance of General Assembly resolutions in international law. They do, and rightly so. The General Assembly is not a representative or law-making body. Each state has one vote; the collective wishes of over 1 billion people in India, expressed through their government's representative in the General Assembly, carry no more voting weight in the General Assembly than the collective wishes of the 290,000 people of Iceland. Most importantly, nothing in the Charter of the United Nations gives the General Assembly any rule-making authority, other than over its own internal housekeeping functions. International law can not be made without the consent of the states affected; as the members of the United Nations have not consented to grant general rule-making authority to the General Assembly, no such consent exists unless it can be shown in some other way, such as through application of the four TOPCO factors. Perhaps surprisingly, these factors have actually been met in some instances involving international environmental law, perhaps most notably in the case of Principle 21 of the Stockholm Declaration on the Human Environment. Portions of this General Assembly resolution, like a few others (such as the Universal Declaration of Human Rights), have acquired a life of their own; their authority comes not from adoption by the General Assembly but from near-universal acceptance, both in the General Assembly vote and the practice of states; from the lack of any objection; and from the lack of any conflicting prior rule.

International environmental law today is derived from a mix of conventional and customary sources; it includes both hard law, creating binding rights and obligations, and aspirational documents. The next part of this chapter traces the development of the regime of international environmental protection from its beginnings to its present form.

Before 1941

The history of modern international law prior to 1941, and especially prior to the first world war, is essentially the history of European international law. Although other regions of the world developed independent traditions of international law as well as of domestic law, these were largely or entirely displaced by European concepts of international law during the age of colonialism. For example, many of the great civilizations of the ancient world arose in relatively water-scarce environments in which irrigation (rather than navigation) was a central focus of economic and political activity (Teclaff, 1991). However, any body of international environmental law emerging from these civilizations was largely occluded by the wholesale exportation of European systems of law during the colonial era.

While the end of colonialism and the emergence of prominent non-European international actors (including, among others, the United States) have greatly lessened European influence, the result of this history is that European ideas and concerns continue to have a disproportionate influence on international law. This European influence on international environmental law is subject to criticism from a variety of angles. From a Third World perspective, it reflects colonialist values; from a U.S. perspective, it reflects an outmoded and nonviable view of global politics; and from a scientific perspective, it fails to take into account the fact that Europe is a small and ecologically atypical continent. Nonetheless, it persists.

In the era prior to the last half of the twentieth century, environmental protection for its own sake was rarely a concern in European international relations. Such protections as are to be found in treaties from this era are incidental to the protection of other interests. Such incidental provisions are particularly likely to be found in treaties dealing with the uses of international watercourses, because there is a nexus between the use and environmental harm. International watercourse treaties are by no means the only area of international rule-making from that era in which incidental environmental protection may be found, but watercourse treaty regimes are particularly comprehensive. They provide valuable examples of the ways in which environmental protection appears either as an afterthought or as an incident to some other end, and the ways in which the environment was subordinated to essentially colonialist goals such as the right to navigation of international waterways within the territory of poorer, less powerful nations. Perhaps more importantly, they demonstrate the evolution of international environmental law from a purely exploitative approach to the first glimmerings of an environmental ethic.

The oldest and most complete river-navigation regime was that governing the Danube river system. In nineteenth-century Europe, rivers were valued more for their navigability and for their potential as sources of power than for the purity of their waters. Because many of Europe's great rivers are international and provide access to the sea for land-locked countries, ensuring freedom of navigation was generally the major focus of treaties regarding these rivers. The colonial powers of Europe, particularly England, France, and Russia, had a strong commercial interest and a less explicit military and intelligence interest in freedom of navigation on the Danube; without it, they had no easy access to the markets and countries of central Europe.

At times the regulation of navigation had environmental consequences. For example, in 1865 the European Commission of the Danube regulated the dumping of ballast and ash into the river, with the fines for the latter being lower. Yet although the dumping of large quantities of inert ballast might have obstructed the flow of water and affected the riparian environment, in the short term it was less likely to affect that environment than was the discharge of chemically active ash. No special provisions were made to account for this potentially harmful environmental effect, as it had little effect on the river's navigability; a river that is biologically dead is just as navigable as a healthy river. The relative importance of navigation and environmental protection is reflected in the fines set for the two types of dumping: fines for unlawfully discharging ash or cinders were set at a lower level than fines for unlawfully discharging ballast (*Public Act of the European Commission of the Danube Relative to the Navigation of the Mouths of the Danube*, 1865).

This distortion was not merely the result of a lack of environmental consciousness in the nineteenth century. Although certainly environmental awareness in the modern sense had not yet arisen, the people of nineteenth-century Romania and Serbia (for example) were capable of recognizing environmental degradation and realizing that they were suffering from it. The distortion was a result of the presence on the European Commission for the Danube of powerful non-riparian nations active in shipping. These states' paramount concern was navigation; they had no interest in other uses of the river, as it lay beyond their territory. In other words, the developmental and environmental needs of the people of Romania and Serbia were of no concern to the representatives of Britain, France, or Russia. They were concerned solely with access for their countries' shipping, not with clean drinking water or plentiful, safe fisheries, nor even with other industrial uses such as the generation of power. In 1881, for example, the European Commission forbade "[t]he establishment in the river, and especially near the banks, of boat-mills, irrigating wheels, and other similar constructions" without authorization from the Commission's river police (*Regulations of Navigation and Police Applicable to the Danube Between Galatz and the Mouths, Drawn up by the European Commission of the Danube*, 1881).

Wood was an important energy source for nineteenth-century river transport, as well as a material for building ships and barges. The Danube treaties anticipated considerable logging along the river and contained provisions to prevent block-age of the navigable channel, but not to control the erosion that this riverside logging was certain to produce. This may have been because the relationship of logging to soil erosion was poorly understood at the time, or it may have been because the erosion was not the non-riparian parties' problem—at least not unless the erosion became severe enough to interfere with navigation. Examples abound of this focus, not only on shipping interests over environment but also on the interests of powerful non-riparians over the interests of those who actually lived along the river. Echoes of this outlook can be seen today in disputes over resources as varied as rain-forest species and satellite orbits.

A more purely environmental concern was the transmission of contagious diseases, although the expression of this concern was somewhat one-sided. The Danube navigation regime exempted vessels in international transit on the river from quarantine and sanitary inspections, except in the event of "contagious pestilence prevailing in the East." The "East" in this case meant the Ottoman Empire; from a twenty-first century perspective it is difficult to say whether this one-sided health measure resulted from actual differences in economic development and public health measures between European countries and the Ottoman Empire or merely from Eurocentric indifference to any effects abroad of pestilence originating within Europe. The most likely explanation is that both factors were present.

World War I ended the old regime of territorial empires in Europe itself. The Austro-Hungarian Empire ceased to exist; Russia and Turkey collapsed from within, the former losing some and the latter much of its territory. Germany also lost territory and was severely restricted by the victorious Allies. In the new world order that emerged, the locus of political, economic, and military power had shifted to the United States. But although some aspects of the regime of international law that emerged after the war reflected this shift, the United Kingdom and France retained their overseas colonial empires and gobbled up much of Germany's as well; the regime governing transboundary watercourses, in particular, continued to reflect the concerns of these two countries to a disproportionate degree.

Gradually the availability of clean, fresh water became more important, while river navigation became less so, both commercially and militarily. The development of the transboundary watercourse regime in the interwar period reflected the development of the fundamental modern concepts of international environmental law, although the culmination of this development would come in a case involving not water pollution but air pollution.

The post–World War I treaties of Versailles and Trianon made extensive provisions for the regime of navigation on the Danube. They reflected a diminished concern with quarantine and safety regulations, but an increasing awareness of the importance of non-navigational uses. Article 293 of the Treaty of Trianon introduced a revolutionary idea: It provided for the creation of a "permanent technical Hydraulic System Commission" having authority over non-navigational uses of the waters of part of the Danube basin. This Commission's scope was specifically environmental: Among other things, it was concerned with deforestation and fisheries. And Article 282 of the Treaty of Trianon provided that, in resolving conflicts, "due allowance" should be made "for all rights in connection with irrigation, water-power, fisheries, and other national interests, which, with the consent of all the riparian States or of all the States represented on the International Commission, shall be given priority over the requirements of navigation." Article 282 accurately anticipated a world in which rivers and other international watercourses would be valued primarily for their non-navigational uses, rather than their navigational uses. (This had also been foreshadowed by the nonbinding 1911 International Regulation Regarding the Use of International Watercourses for Purposes other than Navigation.)

The international regime regarding the uses of transboundary watercourses was crucial to the post–World War I emergence of customary international law regarding transboundary environmental harm. Two decisions of national courts were particularly influential in shaping normative expectations: the 1927 *Donauversinkung* decision and the 1939 *Société Énergie Électrique* decision.

The *Donauversinkung* decision was reached by the Weimar Republic's Staatsgerichtshof (Constitutional Court) in the brief flowering of German democracy between the end of World War I and the rise of the Nazi Party. The case concerned a conflict between two German states, Baden and Wurttemberg, over the waters of the river Aach. (Today Baden-Wurttemberg is a single state.)

In Baden water seeps from the Danube, flows underground, and emerges as the source of the Aach. At the time of the dispute the level of the Danube in neighboring Wurttemberg had fallen significantly as a result of this natural seepage. Wurttemberg sued to prevent Baden from taking any action to increase the rate of seepage. The court held that Baden was prohibited from artificially increasing the rate of seepage, while Wurttemberg was simultaneously prohibited from artificially decreasing it (Lammers, 1984). This essential balancing between the duty of states to do no harm to the territory of other states and the right to utilize resources within one's own boundaries was to become a familiar one in international environmental law.

A more radical concept—the idea that a river basin (or other ecosystem) is a "community"—was expressed in the case of *Société Énergie Électrique du Littoral Mediterraneen v. Compagnie Impresse Elletriche Liguri*. *Société Énergie Électrique* has less respectable political antecedents than the *Donauversinkung* decision; it was decided by a court in Fascist Italy on the eve of the Axis occupation of much of France. The case involved a dispute between two power companies, one in France and one in Italy, over the Italian company's use of the waters of the river Roya. A French court had found in favor of the French power company, but the Italian Corte de Cassazione (court of last appeal) refused to execute the judgment. In dicta (that is, in material not essential to the court's decision in the case), the court stated that "international law recognizes the right of every riparian state to enjoy, as a participant of a kind of partnership created by the river, all the advantages deriving from it. A State cannot disregard the international duty not to impede or to destroy the opportunity of the other States to avail themselves of the flow of water for their own national needs." This "community" theory also flowered briefly in decisions of the Permanent Court of International Justice (PCIJ) prior to World War II, but since then the theory has been kept alive largely by the enthusiasm of scholars and environmental activists rather than by state practice.

During the period between the wars the PCIJ, the judicial organ of the League of Nations and the predecessor of today's International Court of Justice (ICJ), considered three cases involving international watercourses: the 1927 *Advisory Opinion Concerning the Jurisdiction of the European Commission of the Danube Between Galatz and Braila*, the 1929 *Case Concerning the Territorial Jurisdiction of the International Commission of the River Oder*, and the 1937 *Case Concerning the Diversion of Water from the Meuse*. The 1927 *Danube* case gives a thorough and fascinating history of the Danube treaty regime. In the *Meuse* case the PCIJ held that it was limited to the interpretation of an 1863 treaty between Holland and Belgium, the parties to the dispute, and as a result the case is not a useful source of information about the development of international environmental law. (The 1934 *Oscar Chinn* case also involved navigation rights in an international river but is tangential.)

In the *Oder* case the PCIJ gave a more detailed description of the "kind of partnership" mentioned by the Italian Corte de Cassazionne in *Société Énergie Électrique:* There exists, stated the PCIJ, a "community of interest in a navigable river [that] becomes the basis of a common legal right, the essential features of which are the perfect equality of all riparian States in the users [sic] of the whole course of the river and the exclusion of any preferential privilege of any one riparian State in relation to the others." The community theory of international environmental resource management (managing ecological units such as drainage basins as a unit without regard for national boundaries) has obvious advantages from an efficiency standpoint. From the *Oder* case and *Société Énergie Électrique* it can be seen that the community theory was in the air in the late 1930s. But the possibility of such far-reaching international cooperation was lost in the wreckage of World War II.

1941: The *Trail Smelter* Arbitration

The modern international legal regime of state responsibility for transboundary environmental harm is widely considered to have had its genesis in the town of Trail, British Columbia, located not far north of the Canada-U.S. border. The events that ultimately would transform the world's view of transboundary environmental harm and territorial sovereignty began well before 1941. In 1896, a U.S. company built and began operating a smelting plant in Trail to smelt lead and zinc ores; the plant was transferred to Canadian ownership in 1906. In 1925 and 1927, the plant added two 400-foot smokestacks; this, along with an increase in total sulfur emissions at the plant, resulted in an increase in the amount of sulfur dioxide and other fumes reaching the United States.

To the south of Trail, across the border, lies Stevens County, Washington. The plume of sulfur dioxide emissions from the Trail smelter damaged crops, cattle forage, and forests along the Columbia River Valley in Stevens County, from the border to Kettle Falls thirty miles to the south.

The U.S. and Canadian governments twice submitted the dispute to arbitration, once from 1928 to 1931 and a second time from 1935 to 1941. The 1941 decision of the arbitral tribunal set forth *in dicta* the principle that has become the cornerstone of international environmental law:

> No State has the right to use or permit the use of its territory in such a manner as to cause injury by fumes in or to the territory of another or the properties or person therein, when the case is of serious consequence and the injury is established by clear and convincing evidence.

In other words, nations have a responsibility to not allow their territory to be used in ways that cause environmental harm to, or within, the territory of other nations.

The *Trail Smelter* tribunal's holding was binding only on the states involved and only as to the particular issue before the court; that is, it was a definitive statement of the rights and obligations of Canada and the United States in regard to the problem before the tribunal, but it was not a binding precedent in the way that, say, a decision of the U.S. Supreme Court would be a binding precedent on questions of federal law for other courts within the United States.

Furthermore, the oft-quoted paragraph above was not even part of the court's holding; it was dicta (dicta, you will recall, is material not essential to the court's decision in the case.) The *Trail Smelter* decision is perhaps important not so much for its own sake as because it so perfectly captured and enunciated a concept that had been gaining acceptance at least since the time of the *Donauversinkung* case: the idea that each state's territorial sovereignty is limited by a duty not to cause harm to the environments of other states. This concept has since won widespread acceptance in the practice of states undertaken out of a sense of legal obligation and thus may be regarded as a rule of customary international law.

1941–1972: From the *Trail Smelter* Arbitration to the Stockholm Declaration

The Second World War, like the First, brought about a fundamental change in world order. Just as the First World War had ended the old European empires, the Second World War effectively ended the colonial empires of European powers in Africa and Asia, although the actual winding-up took an additional couple of decades (and, in a few corners of the world, longer). Some old concerns evaporated: after World War II, the Treaty of Paris contained a single "Clause Relating to the Danube" governing navigation, in contrast to the Treaty of Trianon's twenty articles on the Danube. The community theory of river basin management was lost in the confusion.

In the postwar era, international tribunals continued to struggle with the idea of limited territorial sovereignty. A few years after the decision of the *Trail Smelter* tribunal, the International Court of Justice (ICJ) expressed the principle in more general terms in the *Corfu Channel* case. In 1946, two British warships struck mines in Albanian waters in the Corfu Channel; many British sailors were killed. Britain first took the matter to the U.N. Security Council, which recommended that the parties submit the dispute to the ICJ, the successor to the PCIJ. In 1949, the ICJ decided that Albania had a duty to warn ships of the mines in the Channel: It is "every State's obligation not to allow knowingly its territory to be used for acts contrary to the rights of other States."

The possibility of community-based approaches to transboundary problems apparent in some cases immediately preceding World War II had thus been replaced with a more traditional sovereignty-based approach: International environmental harm is seen first as an offense against sovereignty, rather than as an offense against the interests of the affected community. The idea of environmental harm as an offense against the environment itself had not even begun to emerge.

Perhaps ultimately even more important to international environmental law was a new theoretical basis for international environmental protection. Although the idea of managing transboundary environmental resources as community resources had been greatly weakened by World War II, a new area of international legal theory began to emerge after the war: the idea of human rights. Modern international human rights law began at Nuremberg, and among the offenses prosecuted at Nuremberg was a purely environmental war crime: Several German officials who had been part of the administration of occupied Poland were charged with "ruthless exploitation of Polish forestry" including "the wholesale cutting of Polish timber to an extent far in excess of what was necessary to preserve the timber resources of the country" (United Nations War Crimes Commission, 1948). As can be seen from the wording, the reasoning here relied on the preexisting Hague regime approach to natural resources: natural resources as property. But as international human rights law developed further, it began to encompass the idea of environmental rights, or of the human right to a healthy environment.

The *Trail Smelter* arbitral tribunal had addressed transboundary pollution *in dicta*. The *Corfu Channel* case had adopted a rule of state responsibility based on the idea of limited territorial sovereignty, but in a non-environmental context. In 1957 an arbitral decision in the *Affaire du Lac Lanoux,* a dispute between Spain and France, clarified this rule of state responsibility and its limits in an environmental context. France proposed to divert the waters of the river Carol to generate electricity. The Carol flows out of the Pyrenees from France into Spain; the diversion was to take place entirely within French territory, and Spain would receive an undiminished flow of the river. Water of the same quality and in the same quantity as that taken from the river would be returned to the Carol before it entered Spanish territory. The arbitral tribunal first expressed the idea that France's territorial sovereignty was limited by its obligation to respect Spain's territorial integrity: "according to the rules of good faith, the upstream state is under the obligation to take into consideration the various interests involved, to seek to give them every satisfaction compatible with the pursuit of its own interests, and to show that in this regard it is genuinely concerned to reconcile the interests of the other riparian State with its own." At the same time, however, Spain's territorial integrity was correspondingly limited by France's territorial sovereignty: "On her side, Spain cannot invoke a right to insist on a development of Lake Lanoux based on the needs of Spanish agriculture. . . . Spain . . . can only urge her interests in order to obtain, within the framework of the scheme decided upon by France, terms which reasonably safeguard them." Ultimately Spain's claim was denied because, as the water reaching Spain was unimpaired in quality and unreduced in quantity, Spain had not shown any harm. To the "do no harm" principle of the *Trail Smelter* and *Corfu Channel* cases, the *Lac Lanoux* decision added a balancing element: States had a right to develop resources within their own borders free from outside interference, so long as that development resulted in no harm beyond the state's borders.

Meanwhile the increasing affluence of the developed countries in the post–World War II era and the culture of consumption continued to put unprecedented strain on the environment. A backlash against the excesses and artificialities of this consumer culture and its apparent determination to conquer or suppress the natural world was inevitable, and in 1962 Rachel Carson published *Silent Spring*, widely credited with launching the modern environmental movement. (Another candidate for this honor is Aldo Leopold's *A Sand County Almanac*, published in 1949.)

Environmentalism was thus an element in the radical political ferment of the 1960s and early 1970s. Among the central concerns of the time was the Vietnam War, which was not without its environmental aspects. In 1967 the U.S. military embarked on what may have been the world's most extreme program of nondevelopment-related environmental destruction: Operation Ranch Hand. The military decided to eliminate large portions of Southeast Asia's forests to prevent their use by the Viet Cong and North Vietnamese Army for concealment. The United States used "Rome plows" (tractors with cutting blades attached) to clear-cut 750,000 acres of land, and seeded clouds in an effort to increase rainfall and render Vietnam's unpaved roads more difficult to use. But the element of Operation Ranch Hand that most captured world imagination and sparked outrage was the aerial spraying of chemical defoliants. Using aircraft, the U.S. forces in Vietnam sprayed 200 million gallons of the herbicides Agent Orange, Agent White, and Agent Blue on Vietnam and Laos. As much as 10 percent of the total land area of South Vietnam may have been sprayed, of which about 86 percent was forest and 14 percent was agricultural land. Napalm and conventional bombing also damaged large areas of forest and agricultural land. These various tactics caused the loss of 8 percent of the region's agricultural land, 14 percent of its forests, and half of its wetlands (Popovic, 1995; Schmitt, 1997; Yuzon, 1996). The environmental damage persists to this day, and people in the affected areas suffer abnormally high rates of birth defects and miscarriages.

1972: The Stockholm Declaration

Growing international environmental awareness generally and Operation Ranch Hand specifically led to a period of environmental treaty-making and customary norm-formation. The environmental movement's concerns for international environmental protection found expression in the first United Nations Conference on the Human Environment, held in Stockholm, Sweden, in 1972.

Representatives of 113 countries took part in the Stockholm Conference. The Soviet Union and its client states in Eastern Europe had originally supported the idea of the conference and participated in the planning. They boycotted the conference itself, however, not because of fundamental disagreements on environmental issues, but to protest the exclusion of the German Democratic Republic (East Germany).

The Stockholm Conference was perhaps most important because it established environmental protection as a distinct goal of both international law and the United Nations process, the latter via the creation of the United Nations Environment Programme (UNEP). It also produced several documents, however; in retrospect the most significant of these has turned out to be the Stockholm Declaration on the Human Environment. The Stockholm Declaration was subsequently adopted by the General Assembly of the United Nations by a vote of 103 for to zero against, with 12 abstentions.

Principle 21 of the 1972 Stockholm Declaration expresses the two strands of thought that we have already seen in the *Donauversinkung, Société Énergie Électrique, Trail Smelter,* and *Corfu Channel* cases: States have the "sovereign right to exploit their own resources pursuant to their own environmental policies." Along with this right, though, comes the responsibility not to cause transboundary harm and, as we shall see, a new element: "responsibility to ensure that activities within their jurisdiction or control do not cause damage to the environment of other States or areas beyond the limits of national jurisdiction." Principle 21 thus extended state responsibility to include harm to areas beyond national jurisdiction: the oceans, the deep seabed, and Antarctica. From an environmentalist perspective, this is only common sense: Important parts of ecosystems lie outside national jurisdiction, and in order to protect an ecosystem as a whole, these parts must also be protected. From an international law perspective, however, the idea of imposing responsibility for actions that did not cause direct harm (or perhaps any harm) to another state was radical. But this too has now become, through state practice, a generally accepted rule of customary international law.

One gap in the protective scope of Principle 21 should be apparent: States now have responsibility for environmental harm caused to areas outside their borders, but they have no duty to avoid environmental harm that takes place entirely within their own borders. Purely domestic harm is outside the scope of customary international environmental law as defined by Principle 21, although it may be the subject of obligations created by international agreements. To this day, purely internal environmental harm remains a problem area for international environmental law; customary international law is unable to address it.

1972–Present: After Stockholm

During the 1970s the pace of international treaty-making continued to increase. In 1976, countries including the United States adopted the Convention on the Prohibition of Military or Any Other Hostile Use of Environmental Modification Techniques (ENMOD), outlawing the hostile use of environmental change such as the deforestation practiced by the United States in Vietnam, as well as possible new forms of environmental modification such as weather control or the deliberate destruction of the ozone layer. And in 1977 the International Committee of the Red Cross promulgated Protocol I to the Geneva Conventions of 1949, which provides in part that "It is prohibited to employ methods or means of warfare which are intended, or may be expected, to cause widespread, long-term, and severe damage to the natural environment." (In contrast, ENMOD prohibits environmental effects that are "widespread, long-lasting, *or* severe.") The United States signed Protocol I in 1978 but has not ratified it; nonetheless, the United States takes the position that much of Protocol I is a statement of binding customary international law.

In addition to ENMOD, over two dozen multilateral environmental treaties were concluded during the 1970s. (The exact number is difficult to pin down because of the difficulty of defining an "environmental" treaty; in the post-Stockholm world, a wide variety of treaties contain environmental provisions.) These treaties cover a broad but not yet complete spectrum of environmental concerns, including hunting, habitat loss, marine pollution, freshwater pollution, air pollution, desertification, and endangered species.

Since the 1970s international treaty-making has continued apace, but while treaty-makers have continued to fill in *lacunae* (gaps) in the treaty regime, customary international environmental law has not advanced significantly in most areas. This is to be expected, as normative expectations about the behavior of states undertaken out of a sense of legal obligation must ordinarily be formed over fairly long periods of time. The very large number of aspirational documents generated in this area of international law creates some confusion, however. The NGOs and intergovernmental organizations (IGOs) involved in producing such documents might uncharitably be described as being in "churn" mode. A number of international environmental law conferences have been held, attended by large numbers of delegates and media representatives, at a measurable environmental cost in jet fuel, waste paper, plastic water bottles, and the other by-products of conferences. The same work might have been done by e-mail (or, in the early days, telephone) with little or no need for conferences, although the media would have been less interested. Perhaps predictably, the world's concern for the environment has spawned a professional class of activists, academics, attorneys, and administrators for whom these conferences serve to justify their professional existence. But the environmental costs of the conferences are negligible when considered in relation to overall global consumption, and the conferences seem to be a necessary part of the gradual process of building customary international law through consensus.

The World Charter for Nature, adopted by the U.N. General Assembly in 1982 by a vote of 111 in favor to 1 (the United States) against, might be expected from the vote alone to be a radical document from a U.S. viewpoint. (Algeria, Lebanon, and several Latin American countries abstained.) Yet it retains what by 1982 had already become the traditional approach to questions of state responsibility for transboundary harm: "States . . . shall . . . [e]nsure that activities within their jurisdictions or control do not cause damage to the natural systems located within other States or in the areas beyond the limits of national jurisdiction[.]" This is identical to, and presumably drawn from, the language of

Principle 21 of the 1972 Stockholm Declaration. The World Charter for Nature also incorporates Principle 21's recognition of sovereign development rights: States have the "sovereign right to exploit their own resources pursuant to their own environmental policies[.]"

The Rio and Johannesburg conferences have not altered the fundamental concepts expressed at Stockholm. The norm-formation process that is customary international lawmaking proceeds slowly, and it may be many years before any changes wrought at Rio or Johannesburg win widespread acceptance in state practice. Meanwhile the so-called North-South conflict that was apparent at Stockholm has emerged as the dominant tension in international environmental customary norm-formation and, to a lesser extent, treaty-making.

The 1992 Rio de Janeiro Conference on Environment and Development, like the Stockholm Conference before it, produced a declaration. Principle 2 of the Rio Declaration is nearly identical to Principle 21 of the Stockholm Declaration:

> States have, in accordance with the Charter of the United Nations and the principles of international law, the sovereign right to exploit their own resources pursuant to their own environmental *and developmental* policies, and the responsibility to ensure that activities within their jurisdiction or control do not cause damage to the environment of other States or areas beyond the limits of national jurisdiction. [Emphasis added]

The sole difference is the two words "and developmental," added to reflect the concerns of developing countries and the growing effectiveness of those countries at making their voices heard. While it may indicate a shifting of the balance in the North-South controversy, it adds no new dimension to a conflict that was already clearly in evidence at Stockholm.

The Rio Declaration is not a mere restatement of the Stockholm Declaration, however. It incorporates several ideas present in the post-Stockholm environmental discourse. It places more stress on the idea of intergenerational equity: "[T]he right to development must be fulfilled so as to equitably meet developmental and environmental needs of future generations." The Rio Declaration also incorporates the polluter pays principle: "National authorities should endeavor to promote the internalization of environmental costs . . . the polluter should, in principle, bear the cost of pollution[.]" And it incorporates the precautionary principle: "In order to protect the environment, the precautionary approach shall be widely applied by States according to their capabilities. Where there are threats of serious or irreversible damage, lack of full scientific certainty shall not be used as a reason for postponing cost-effective measures to prevent environmental degradation."

The Rio Declaration also disagrees with the *Lac Lanoux* tribunal's decision that no notice of environmentally significant projects that may have a transboundary effect must be given to potentially affected states: The Declaration requires environmental impact assessment and "prior and timely notification . . . to affected states[.]" This is, of course, purely aspirational; it has not yet become a norm of customary international law evidenced by state practice undertaken out of a sense of legal obligation.

The declaration adopted by the World Summit on Sustainable Development in Johannesburg, South Africa, in 2002 takes a different approach; it focuses on the burdens borne by developing countries in the attempt to achieve sustainable development. At this early point, at least, it is perhaps best viewed as an expression of concern rather than as part of any process of international norm-formation.

In other words, customary international environmental law for the most part adheres to the principles established in or by the time of the Stockholm Declaration, although more recent state practice may affect the interpretation of those principles.

It should be emphasized that none of these aspirational documents represents "law," except to the extent that they reiterate principles already generally accepted in state practice. None has gained the acceptance in the practice of states, undertaken out of a sense of legal obligation, necessary for the formation of normative expectations. At worst they are feel-good documents without any impact whatsoever; at best they express aspirations that may some day come to be accepted as norms of customary international law, but they have not yet attained that status.

On the conventional (treaty) side, a limited but dramatic development in recent years has been the emergence of the International Criminal Court (ICC). The organic document of the ICC is the Rome Statute, article 8(2)(b)(iv) of which prohibits

[i]ntentionally launching an attack in the knowledge that such attack will cause incidental loss of life or injury to civilians or damage to civilian objects or widespread, long-term, and severe damage to the natural environment which would be clearly excessive in relation to the concrete and direct overall military advantage anticipated.

(Note the similarity of Protocol I to the Geneva Conventions, including the use of the word "and" in place of ENMOD's "or.") Article 8(2)(b)(iv) thus gives the ICC jurisdiction over some environmental war crimes. This may be only a tiny percentage of international environmental crimes, let alone of international environmental harm caused by noncriminal negligence or recklessness, but it is at least a beginning: For the first time a permanently constituted international tribunal has specific and unequivocal jurisdiction over a category of international environmental harm. (The United States, however, is not a party to the Rome Statute.)

The Role of the United Nations and Its Specialized Agencies

The involvement of the United Nations in international environmental norm-formation, enforcement, and dispute resolution is complex. Almost every organ and agency of the United Nations may perform functions, at least occasionally, with environmental significance. Fortunately, the environmental activities of the various agencies of the United Nations are monitored and coordinated by the United Nations Environment Program (UNEP).

UNEP was created as a result of the Stockholm Conference in 1972, not by treaty but by Resolution 2997 of the U.N. General Assembly. This was an action within the General Assembly's housekeeping authority, as UNEP is merely a clearinghouse; it has no executive powers of its own (other than over its own internal housekeeping functions). Instead UNEP serves to prevent the many agencies of the United Nations with environmental functions from needlessly duplicating each other's work or working at cross-purposes and provides easier access to environmental information. All UNEP programs are financed directly by member states; UNEP coordinates the environmental activities and information-gathering of other U.N. agencies, including:

- Commission on Sustainable Development (CSD)
- Food and Agriculture Organization (FAO)
- International Atomic Energy Agency (IAEA)
- International Labor Organization (ILO)
- International Maritime Organization (IMO)
- United Nations Conference on Trade and Development (UNCTAD)
- United Nations Development Program (UNDP)
- United Nations Educational, Scientific and Cultural Organization (UNESCO)
- United Nations Institute for Training and Research (UNITAR)
- World Health Organization (WHO)
- World Meteorological Organization (WMO)

UNEP is an indispensable resource for anyone working in the field of international environmental law; contact information is provided in Chapter 7.

In addition to UNEP and the agencies listed above, other U.N. entities that merit particular attention are the Global Environment Facility (GEF), the World Bank and the International Monetary Fund (IMF), the International Law Commission (ILC), and the ICJ.

The GEF, established in 1991, provides financial assistance to programs in developing countries that protect the global environment. The GEF can provide grants to support projects related to obligations under international treaties on biodiversity and climate change, as well as projects to protect international waters and the ozone layer, prevent land degradation, and reduce and control persistent organic pollutants (Guruswamy, 1997).

The World Bank (actually a group of five organizations: the International Bank for Reconstruction and Development, the International Development Association, the International Finance Corporation, the Multilateral Investment Guarantee Agency, and the International Centre for Settlement of Investment Disputes) and similar institutions, such as the European Bank for Reconstruction and Development, provide loans for development projects. The World Bank's importance is magnified by the fact that many private lenders follow the World Bank's lead in deciding whether to loan money for a particular development project. The IMF is principally responsible for ensuring international financial stability; its actions can often affect the policies of developing-country governments, with environmental consequences (Guruswamy, 1997).

Until recently both the World Bank and IMF were considered unequivocally bad by environmentalists; both had a record of encouraging unsound development projects and policies and of valuing short-term returns at the cost of long-term economic harm. Under pressure from environmental activists and donor governments, both have now, at least on the surface, changed their tune: the IMF states that "Only sustainable economic growth—a central aim of the IMF's policy advice—can generate the additional resources needed to address environmental problems" (International Monetary Fund, 2004). The actual meaning of this statement, if any, is somewhat ambiguous: Does the IMF mean that economic growth must be environmentally sustainable, or that only rich countries can afford to address environmental problems? For its part, the World Bank in 2001 released a document titled "Making Sustainable Commitments: An Environment Strategy for the World Bank" that makes a similarly ambivalent commitment to sustainable development. Despite these public statements and evidence of increased environmental sensitivity in investment decisions, the World Bank and IMF remain a cause for concern because of the opacity of their decision-making processes.

The International Law Commission is an entity created by the General Assembly to attempt to bring about a codification of international law; projects of particular importance to international environmental law have been the ILC's Draft Articles on State Responsibility and on the Non-navigational Uses of International Watercourses; the latter, with modifications, was adopted in 1997 by the General Assembly as the United Nations Convention on the Law of the Non-navigational Uses of International Watercourses, now ratified or otherwise accepted by twelve countries but not yet in force.

The ICJ is the United Nation's judicial organ; it has jurisdiction over those states that have consented to its jurisdiction either generally or for a specific matter. The full court has addressed major international environmental disputes such as those arising from the use of nuclear weapons or the construction of a large hydroelectric plant along the border between Slovakia and Hungary. In 1993, the ICJ created a Chamber for Environmental Matters to streamline the resolution of environmental disputes; however, no cases have ever actually been brought before the chamber. Perhaps significantly, the aforementioned suit between Slovakia and Hungary was brought before the entire ICJ by special agreement between the parties, although the parties could as easily have chosen to submit the dispute to the Chamber for Environmental Matters. In 2001, the Permanent Court of Arbitration, another major international dispute-resolution body, adopted its Optional Rules for Arbitration of Disputes Relating to Natural Resources and/or the Environment; these rules were designed to address some of the problems, such as lack of particular environmental expertise on the part of the judges assigned to the panel, that may have led to the nonuse of the ICJ's environmental chamber.

Fundamental Concepts in International Environmental Law

Rule-Making, Enforcement, and Dispute Resolution

Any regime of control, whether municipal (domestic) or international, must perform three functions: rule-making, enforcement, and dispute resolution. Rules of customary international law are normative expectations arising from the practice of states undertaken out of a sense of legal obligation. The customary norm-formation practice ordinarily takes place slowly. In the preceding portion of the chapter we have looked at state practice giving rise to normative expectations regarding state responsibility for transboundary environmental harm; this section examines the rule-making process under multilateral environmental agreement regimes.

Rule-Making. In the years before the 1972 Stockholm Conference, those multilateral environmental protection regimes that came into existence tended to be in the form of attempts at a one-time solution to a perceived problem. In the post-Stockholm world the rule-making process has grown more sophisticated. Multilateral environmental protection regimes typically begin with an initial framework treaty, with relatively little infringement on the sovereignty of the parties. The framework treaty sets the stage for a Conference of the Parties (COP) to adopt further protocols on the same topic. This framework-treaty-plus-protocol process makes it possible to reach agreements in a series of small steps, when one large step might have been impossible.

The process set forth in the Vienna Ozone Convention is illustrative: The convention itself amounts, substantively, to little more than an agreement to take "appropriate measures" for the protection of "human health and the environment against adverse effects resulting or likely to result from human activities which modify or are likely to modify the Ozone Layer[.]" The purpose of the convention was to serve as a starting point: Article 6 of the convention provides for an annual meeting of a COP. Articles 2 and 8 provide that the COP may adopt protocols (new treaties binding only on those states that consent) to further the purpose of the convention. Article 9 provides that the convention may be amended by three-fourths vote of the COP; amendments thus adopted must subsequently be ratified, approved, or accepted by the parties and thus need not bind dissenting parties. The same procedure applies to the amendment of protocols, except that only a two-thirds vote is required. Although this would seem absurdly cumbersome as a rule-making mechanism in domestic law, it is an exceptionally effective one from an international law perspective.

In 1987 the Vienna Ozone Convention's COP adopted the Montreal Protocol, which itself has been adjusted or amended in 1990, 1992, 1995, 1997, and 1999. From an initial weak framework convention, the ozone treaty and its protocol and amendments have grown incrementally into a strong and effective environmental protection regime.

The one-time solution process presented both physical and political problems. The physical dimensions of the problem might change over time, or advances in scientific understanding of the problem might reveal it as more or less serious than originally thought. Technological advances might present new, simpler solutions to the problem that were previously unavailable. A change in consumer or producer behavior might suddenly render a major problem minor, or cause a previously ignored aspect of the problem to become crucial.

These problems with the one-time solution approach, significant as they were, were nonetheless outweighed by the political difficulties of such an approach. Participation in a multilateral environmental protection regime inevitably involves the acceptance of limitations on sovereignty by the states involved; they sign away the right to do something that previously it was legal for them to do. With a one-time solution, states must sign away a large packet of sovereign rights all at once, and they are naturally reluctant to do so. This means either that some states will refuse to participate or that the regime must be watered down to attract more states, thus becoming less effective.

The framework-plus-protocol approach acknowledges this difficulty; instead it seeks first to have states agree that a particular goal (such as minimizing the effects of ozone depletion) is desirable. Once this agreement has been reached in principle, states are asked to undertake some activities, such as scientific research and information-sharing, to which few are likely to object. They are also asked to meet again after they have gathered and shared the information and to reach such new agreements as may be necessary.

This approach recognizes that, while states may not be able to resolve environmental problems through a single negotiation, incremental agreement is possible. The incremental agreement is an easier sell: States can perceive the benefits of prior stages of the agreements; they can see that progress has been made in resolving the underlying environmental problem; they can see that other parties have made similar sacrifices and acted in good faith; and the incremental infringement on sovereignty may be perceived as less painful than an "all at once" loss would be.

Enforcement.   Enforcement is, from the point of view of lawyers accustomed to domestic legal systems, the weak point of international law. International agreements partake more of the nature of contracts than of statutes, despite superficial similarities to the latter. In the event of breach, there is in most cases no international or supranational institution with authority to enforce compliance. (In a few treaty regimes, such an institution has been specifically created by the treaty itself, or some already-existing entity may have been endowed with such powers.) Instead, compliance relies to a large extent on the parties having bargained in good faith and acting in good faith to fulfill their obligations. Noncompliance may result in harm to the reputation of the non-complying country and a consequent reluctance on the part of other countries to trust that country in the future. Treaty regimes may also impose sanctions for noncompliance or provide for the suspension or expulsion of noncomplying countries from the treaty organization.

Dispute resolution. A variety of dispute-resolution mechanisms are available in international law. Article 33 of the United Nations Convention on the Law of the Non-navigational Uses of International Watercourses (not in force) provides a fairly comprehensive list:

1. Negotiation;
2. Good offices, conciliation, mediation, or submission of the dispute to a joint . . . institution, arbitration, or the ICJ;
3. Appointment of an impartial fact-finder at the request of any party to the dispute after six months without resolution through one of these means. The fact-finder will make recommendations, by majority vote, which the parties are obligated to consider in good faith.

Negotiation is the way most disputes, domestically or internationally, tend to be resolved; the introduction of a third party into the process is necessary only when negotiation fails. The terms "good offices," "mediation," and "conciliation" create roughly similar impressions in their ordinary English usages. Each, however, has a specific and distinct meaning in international law. A person, state, or entity that lends its good offices to the resolution of a dispute, or agrees to mediate, acts to enable parties that have become estranged to communicate with each other indirectly. The 1899 Hague Convention on the Pacific Settlement of Disputes provides that the role of the mediator is to reconcile "the opposing claims and [to appease] the feelings of resentment that have arisen between the States at variance." Good offices, in contrast, have been described as "tending to call negotiations between conflicting states into existence," rather than making proposals for the states to consider and negotiate upon, as a mediator does. Conciliation is a process by which a permanent or ad hoc body appointed by the parties can inquire into facts and law and make recommendations. Conciliation is a nonbinding attempt by a third party to define terms of settlement between parties who find themselves unable to do so; it may be thought of as a highly formalized type of mediation

(Merrills, 1991). The process of the ICJ and of arbitral tribunals resembles in general outline the process of their domestic-law analogues—domestic litigation and arbitration.

The third category in Article 33's list is known more concisely as "inquiry." Inquiry is conducted by an ad hoc international panel (commission of inquiry) that looks only or primarily at questions of fact (rather than law). Inquiry as a dispute resolution mechanism has its origins in the failure to satisfactorily resolve the Maine incident, which led to the Spanish-American War. It is embodied in the 1899 and 1907 Hague Conventions, but has been little used in modern times (Merrills, 1991).

The "North-South" Conflict: The Underlying Tension between Environment and Development

The fundamental tension in international, or for that matter domestic, environmental law is between environmental values and development. It is an oversimplification to say that rich countries are lined up on the "environmentalist" side while developing countries make development a higher priority, but there is certainly a widespread belief in developing countries that the environmentalism of the developed countries is hypocritical and harmful to the interests of the developing countries (D'Amato and Engel, 1996). This is sometimes expressed in the saying, "Sustainable development means no development."

This conflict of interests is often referred to as the "North-South conflict." This may be a useful shorthand, so long as it is understood that it does not necessarily refer to the actual geographic location of all of the countries concerned. Australia, for example, is physically in the south but economically, culturally, and politically solidly within the developed north, despite the assertions of some Australian leftists to the contrary. Greenland, on the other hand, lies far to the north, yet is still at a level of development that makes its concerns those of the south. In the case of Greenland, matters are complicated yet further by colonial rule—while Greenland is internally self-governing, its foreign affairs are under the control of developed, "northern" Denmark (Turner, 2001).

"North-South conflict" must also be understood as a generalization; while countries grouped in the "northern" or "southern" groups may have common interests with others in the same group, those interests will not prevent countries in the same group from coming into conflict with each other. Countries that share a border will inevitably come into conflict over the use of transboundary resources or as a result of transboundary pollution. The United States is equally likely to become embroiled in environmental disagreements with developed Canada or with developing Mexico.

And development levels are relative. Landlocked, "southern" Mongolia may, by virtue of its geography and level of development, take an approach to international environmental issues similar to that of equally landlocked, truly southern Lesotho. Both countries are intermittently involved in disputes with larger, more developed neighbors over such things as water resources: Russia and China in the case of Mongolia, South Africa in the case of Lesotho. While these neighbors might credibly claim to be developing nations and thus part of the global "south," from the point of view of Mongolia and Lesotho they represent the "north."

The developing countries' mistrust of the developed world's environmental agenda is not merely a legacy of the colonial era; it has very real and well-founded roots. The policies of northern countries on a variety of matters, from agricultural policy to the placement of communications satellites and the development of new pharmaceuticals, can reasonably be interpreted as motivated by a desire to keep the southern countries poor. The northern countries have no such malicious intent, of course, any more than Britain and France intended to deny clean drinking water to the people of the lower Danube Basin in the early days of the Danube navigation regime. What Britain and France wanted was to ensure freedom of navigation for their shipping; they were not actively hostile to interests that might be harmed in achieving that goal or that were inconsistent with it. They were simply indifferent; and today's global north is not actively hostile to the development interests of the South, but again simply indifferent. The rich countries of the world want to protect their farmers, build telecommunications networks, and patent and market pharmaceuticals; in doing so they often trespass on the development of interests of the South, often without being aware that they are doing so. This indifferent trampling of southern interests gives rise to some of the disputes that will be examined in the next chapter, as well as to a generalized climate of southern mistrust of environmental proposals emanating from the developed North.

Sustainable Development and Related Concepts

Underlying all modern concepts of international environmental law is the principle of limited territorial sovereignty. Limited territorial sovereignty is an inevitable consequence of the customary international law concept of state responsibility developed through the practice of states undertaken out of a sense of legal obligation and expressed in the *Trail Smelter* arbitration, the *Corfu Channel* case, the *Lac Lanoux* arbitration, the Stockholm Declaration, and numerous other documents. Limited territorial sovereignty restricts the development that may take place within the borders of a nation to that which causes no undue harm outside those borders. Recently, however, the idea that development must also be sustainable has increasingly been voiced.

A central tension in the development of U.S. environmental law, and perhaps consequently in the development of international environmental law as well, has been the tension between preservationist and conservationist approaches to environmental protection. The preservationist goal is to protect the environment by limiting the exploitation of natural resources as much as possible. For many preservationists this approach is underlain by a belief, more spiritual than economic in nature, that the natural environment has value in its unexploited state, not for any tangible benefit it may bring to humanity, but for its own sake. In contrast, the conservationist goal is "wise use"—the responsible, limited use of natural resources. The tension between these two approaches is uneasily balanced in the concept of sustainable development.

The 1987 U.N.-commissioned Brundtland Report defined "sustainable development" as "development which meets the needs of the present generation without compromising the ability of future generations to meet their needs." The question of whether sustainable development as a whole is now a guiding principle of customary international law is debatable. The ICJ stated in the *Gabčíkovo-Nagymaros* case that sustainable development is a sociopolitical objective, not a binding norm of customary international law. However, in a separate opinion much cited by advocates of sustainable development, the ICJ's vice president, Judge Weeramantry, stated that sustainable development has received wide and general acceptance by the global community and is a principle of customary international law. Even if sustainable development as a whole is not now required by customary international law, certain component concepts within the broad category of ideas collected under the heading of "sustainable development" may and certainly have generated considerable discussion. The Rio Declaration sets out twenty-seven principles to guide the international community in achieving sustainable development, including the "polluter-pays" principle, the "precautionary" principle, and the principle of "intergenerational equity."

The polluter-pays principle: A central problem for modern environmental law is the internalization of the traditionally external costs of environmental harm. Disposing of wastes costs money; wastes improperly disposed of cause harm to the environment and thus to persons other than the polluter. The cost savings to the polluter (who does not pay the costs of proper disposal) are borne by the polluter's neighbors (in the form of environmental harm). This is an example of Garrett Hardin's famous "tragedy of the commons." If costs are not internalized, there is no economic incentive to avoid doing environmental harm; in fact, there is an incentive to engage in environmentally destructive behavior. Using a "pasture open to all" as his example, Hardin explains that:

> It is to be expected that each herdsman will try to keep as many cattle as possible on the commons. . . . He asks, "What is the utility to me of adding one more animal to my herd?" This utility has one negative and one positive component. The positive component benefits the herdsman alone, while the negative component is shared equally by all of the herdsmen. Thus, as long as there is more than one herdsman, it will always be to his individual benefit to over-exploit the commons.

Although the polluter-pays principle is a rare example of an idea that makes good sense from either a "green" or an economic approach to environmental protection, it is not clear that it has yet become a general rule of customary international law through incorporation into the practice of states out of a sense of legal obligation. It has, however, found its way into numerous international agreements and aspirational documents, including the United Nations Convention on the Protection and Use of Transboundary Watercourses and International Lakes (art. 2.5(b)), the Rio Declaration (Principle 16), Agenda 21 (§§18.15 and 18.40(b)(i)), and the Johannesburg Plan of Implementation (para. 19(b)).

The precautionary principle: Some forms of environmental damage are potentially so severe that measures that might prevent them may have to be taken even before the efficacy of and necessity for those measures can be determined. Although a precise definition of the precautionary principle is difficult to pin down, a useful starting point can be found in the second sentence of Principle 15 of the Rio Declaration: "Where there are threats of serious or irreversible damage, lack of full scientific certainty shall not be used as a reason for postponing cost-effective measures to prevent environmental degradation." Others would prefer a stronger statement of the principle, resulting in a more preservationist approach. For example, the Wingspread Statement on the Precautionary Principle, an aspirational document less well-known than the Rio Declaration, states the principle quite a bit more broadly. (The odd name of the Wingspread Statement comes from the Wingspread Conference Center in Racine, Wisconsin, designed by the celebrated architect Frank Lloyd Wright.) The statement dispenses with the "serious or irreversible damage" requirement in Principle 15: "When an activity raises threats of harm to human health or the environment, precautionary measures should be taken even if some cause-and-effect relationships are not fully established scientifically." This broader statement would seem to be almost impossible to put into practice.

The precautionary principle is subject to attack on economic grounds: It may lead to inefficient allocation of resources. In the worst case, of course, the perception of harm may ultimately turn out to have been incorrect, and a great deal of money, time, and effort may have been expended unnecessarily. Nonetheless, when the magnitude of the potential harm rises to the level of a global catastrophe (such as climate change or ozone depletion), precautionary measures in advance of conclusive scientific evidence may be necessary.

Nonetheless, various formulations of the precautionary principle have found acceptance in international agreements and aspirational documents, and some theorists argue that it has become a principle of customary international law, through acceptance in the practice of states undertaken out of a sense of legal obligation. Documents incorporating some form of the precautionary principle include the Rio Declaration (Principle 15), the Cartagena Protocol on Biosafety to the Convention on Biological Diversity (preamble), the World Trade Organization Agreement on the Application of Sanitary and Phytosanitary Measures (art. 5.7), the Stockholm Convention on Implementing International Action on Certain Organic Pollutants (preamble and arts. 1, 8, and 9), the European Union's draft constitution (art. III-129.2), and the United Nations Framework Convention on Climate Change (UNFCCC) (art. 3[3]). The ozone depletion treaty regime generally—the Vienna Ozone Convention and the Montreal Protocol—was the precautionary principle's finest hour. Precautionary measures were formally required by the treaty regime at a time when the scientific necessity for those measures had not yet been established. Later scientific evidence, however, has revealed both that the danger was real and that the protective measures instituted in the treaty regime were necessary and, apparently, effective (UNEP Ozone Secretariat, 2000).

Intergenerational equity: Intergenerational equity is far more controversial than the polluter-pays principle, and perhaps more controversial than the precautionary principle. The concept of intergenerational equity is inherent in the Brundtland Report's definition of sustainable development as development that meets the needs of present generations while not compromising the ability of future generations to meet their own needs. This is often expressed by environmentalists in terms along the lines of "we hold the Earth in trust for future generations" (Weiss et al., 1998). As a result, we have a duty not to waste the principal in that trust by committing environmental damage or exploiting resources unsustainably.

The idea is subject to attack from two directions. One is the idea that the Earth belongs to the living, sometimes expressed in the question "What have future generations ever done for us?" A second, related attack is that "intergenerational equity" is yet another mechanism for holding environmental resources inviolate—a preservationist approach to environmental protection that keeps the poor countries poor and the rich countries rich. The true duty of the present inhabitants of poor countries to their descendants may lie, not in passing along to them an unaltered natural environment, but in becoming wealthy so that their descendants may stand on an equal footing with the inhabitants of the global north.

It cannot be said that intergenerational equity has found widespread acceptance in the practice of states undertaken out of a sense of legal obligation. Thus, it has not attained the status of a norm of customary international law. It has, however, been incorporated into a number of international agreements and aspirational documents, including the Rio Declaration (Principle 3), the Non-Binding Authoritative Statement of Principles for a Global Consensus on the Management, Conservation and Sustainable Development of all Types of Forests (Principle 2(b)), the Convention on Biological Diversity (CBD) (preamble), and the UNFCCC (preamble and art. 3(1), Agenda 21 (art. 8.7).

Philosophical Approaches to International Environmental Law

Theorists building on the philosophical foundations laid by Rachel Carson, Aldo Leopold, and others have constructed complex systems of environmental ethics, but the guiding ethical principles of international environmental law remain rudimentary. The earliest approach was a sovereign rights approach: the environment had value only insofar as it was under the authority of a sovereign state. This approach had the virtue of rendering concerns of preservation versus conservation irrelevant on the international stage: A sovereign state was free either to preserve environmental resources within its territory untouched or to treat them as an economic resource, and in either case any interference with the state's chosen course of conduct was an infringement on its sovereign rights. This sovereign rights approach is still the main ethical underpinning of international environmental law.

The second approach, which has gradually emerged since World War II but has only recently begun to gain acceptance, is a human rights approach: the environment has value because human beings depend upon it (Sachs, 1995). A third approach is the "deep green" approach: the natural environment has value for its own sake, and it is wrong for human beings to infringe upon it (Sessions, 1995). While this theoretical approach might seem to be reflected in international environmental law dealing with some parts of the global commons—the Antarctic treaty regime and Principle 21's extension of the duty not to cause environmental harm to areas beyond national jurisdiction—a more probable theoretical underpinning for these examples is that the areas protected are the common heritage of all humankind and thus protected on the basis of some sort of combination of sovereign and human rights, rather than for their own sake.

The idea of green rights as human rights has been growing for some time, though, and has found acceptance in numerous international agreements and aspirational documents. It appears in the Stockholm and Rio declarations, and in dozens of newer national constitutions such as those of Hungary and South Africa. In conventional international law the connection between human rights and environmental destruction in wartime can be seen in Protocol I to the Geneva Conventions and to a lesser extent in Protocol II, and references to environmental rights appear in the Convention on the Rights of the Child, the African Charter on Human and Peoples' Rights, and the Additional Protocol to the American Convention on Human Rights.

Summary

This chapter provides an overview of the development of international environmental law, providing a historical background that covers the past century with particular attention to events since 1941. It includes information about the sources of international law, the development of customary and conventional international environmental law, and the role of the United Nations and its specialized agencies in international environmental law. It describes the framework convention/conference of parties/supplemental protocol rule-making process common to modern multilateral environmental agreement regimes and looks at emerging principles and underlying philosophical concepts in international environmental law. It also provides an introduction to the "North-South" conflict: the inevitable tension between environment and development, and consequently between developed and developing countries, that underlies many if not most disputes in environmental law.

International environmental law had no separate existence prior to the 1941 *Trail Smelter* decision; where environmental protection existed in a regime of international law, it was generally incidental to some other goal such as the protection of navigable waterways or dealt with as an economic interest in some specific resource, such as fur seals. From the *Trail Smelter* decision came the idea of state responsibility for environmental harm beyond the state's borders; this came to be generally accepted in state practice and was embodied in Principle 21 of the Stockholm Declaration. Since the Stockholm Declaration, many multilateral environmental agreement regimes have emerged, so that most forms of environmental harm are addressed at least to some degree in international environmental law.

After World War II a parallel strand of international environmental legal thought also emerged: the idea of environmental rights as human rights. This idea has been slower to gain widespread acceptance but has been embodied in national constitutions and some international agreements.

The idea of sustainable development and related concepts, including the precautionary principle and intergenerational equity, have been controversial because they are seen as pitting the interests of the developed "North" against those of the developing "South." Many, although not all, international environmental disputes arise from this conflict; resolving it is perhaps the greatest challenge presently facing international environmental law.

Sources and Further Reading

Books and Articles

Abbey, Edward. *Desert Solitaire: A Season in the Wilderness*. New York: McGraw Hill, 1968.

Bales, Jennifer S. *Transnational Responsibility and Recourse for Ozone Depletion*. 19 Boston College International and Comparative Law Review 259 (1996).

Bederman, David J. *International Law Frameworks*. New York: Foundation Press, 2001.

Benvenisti, Eyal. *Sharing Transboundary Resources: International Law and Optimal Resource Use*. Cambridge, UK: Cambridge University Press, 2002.

Bergesen, Helge Ole, et al., eds. *The Fridtjof Nansen Institute Yearbook of International Co-operation on Environment and Development, 7th ed.* London: Earthscan Publications, 1998.

Boer, Ben, et al. *International Environmental Law in the Asia Pacific.* The Hague: Walter Kluwer, 1998.

Bolla, Alexander J., and Ted L. McDorman, eds. *Comparative Asian Environmental Law Anthology.* Durham, NC: Carolina Academic Press, 1999.

Brown, Lester R. et al. *State of the World 2003: A Worldwatch Institute Report on Progress Toward a Sustainable Society.* New York: W.W. Norton, 2003.

Buergenthal, Thomas, and Sean D. Murphy. *Public International Law in a Nutshell, 3rd ed.* St. Paul, MN: West, 2002.

Caron, David. *The Frog That Wouldn't Leap: The International Law Commission and Its Work on International Watercourses.* 3 Colorado Journal of International Environmental Law and Policy 269 (1992).

Carson, Rachel. *Silent Spring.* Boston: Houghton Mifflin, 1962.

Cassese, Antonio. *International Law.* Oxford: Oxford University Press, 2001.

Costa, Pascale. *Les effets de la guerre sur les traites relatifs au Danube, dans le cadre d'une etude globale du droit conventionnel du Danube,* in Ralph Zacklin and Lucius Caflisch, eds., *The Legal Regime of International Rivers and Lakes/Le regime juridique des fleuves et des lacs internationaux.* Dordrecht, Netherlands: Martinus Nijhoff, 1981, 203.

D'Amato, Anthony, and Kirsten Engel, eds. *International Environmental Law Anthology.* Cincinnati: Anderson Publishing, 1996.

Dellapenna, Joseph W. *Treaties as Instruments for Managing Internationally-Shared Water Resources: Restricted Sovereignty v. Community of Property.* 26 Case Western Reserve Journal of International Law 27 (1994).

Dixon, John A., et al. *Economic Analysis of Environmental Impacts, 2nd ed.* London: Earthscan Publications, 1994.

Franck, Thomas M. *Fairness in International Law and Institutions.* Oxford: Oxford University Press, 1995.

French, Hilary F. *After the Earth Summit: The Future of Environmental Governance.* Washington, DC: Worldwatch Paper No. 107, 1992.

————. *Partnership for the Planet: An Environmental Agenda for the United Nations.* Washington, DC: Worldwatch Paper No. 126, 1995.

Garibaldi, David Ray, and Richard Falk, eds. *Postmodern Politics for a Planet in Crisis: Policy, Process, and Presidential Vision.* Albany: SUNY Press, 1993.

Garibaldi, Oscar M. *The Legal Status of General Assembly Resolutions: Some Conceptual Observations*, Proceedings of the American Society of International Law 324 (1979).

Guruswamy, Lakshman. *International Environmental Law in a Nutshell.* St. Paul, MN: West, 1997.

Hardin, Garret. *The Tragedy of the Commons.* 162 Science 1243 (1968).

Janis, Mark W. *An Introduction to International Law, 3rd ed.* New York: Aspen, 1999.

Johnson, Stanley P., and Guy Corcelle. *The Environmental Policy of the European Communities.* The Hague: Walter Kluwer, 1995.

Khadduri, Majid. *The Islamic Law of Nations.* Baltimore: Johns Hopkins Press, 1966.

Kibel, Paul Stanton. *The Earth on Trial: Environmental Law on the International Stage.* New York: Routledge, 1999.

Kiss, Alexandre, and Dinah Shelton. *Manual of European Environmental Law.* Cambridge, UK: Grotius, 1993.

Knowledge for Sustainable Development: An Insight into the Encyclopedia of Life Support Systems, 3 vols. Paris: UNESCO Publishing, 2002.

Lammers, Johan G. *Pollution of International Watercourses.* Dordrecht, Netherlands: Martinus Nijhoff, 1984.

Leopold, Aldo A. *Sand County Almanac.* New York: Oxford University Press, 1949.

Logan, Bruce E. *Environmental Transport Processes.* New York: John Wiley and Sons, 1999.

Lomborg, Bjørn. *The Skeptical Environmentalist: Measuring the Real State of the World.* Cambridge, UK: Cambridge University Press, 2001.

Masters, Suzette Brooks. *Environmentally Induced Migration: Beyond a Culture of Reaction.* 14 Georgetown Immigration Law Journal 855 (2000).

McDougal, Myres S. *Contemporary Views on the Sources of International Law: The Effect of U.N. Resolutions on Emerging Legal Norms,* Proceedings of the American Society of International Law 327 (1979).

McNeill, J. R. *Something New Under the Sun: An Environmental History of the Twentieth-Century World.* New York: W.W. Norton, 2000.

Merrills, John G. *International Dispute Settlement.* Cambridge, UK: Grotius, 1991.

O'Riordan, Timothy, and Heather Voisey. *The Transition to Sustainability: The Politics of Agenda 21 in Europe.* London: Earthscan Publications, 1998.

Plater, Zygmunt B., et al. *Environmental Law and Policy: Nature, Law, and Society.* St. Paul, MN: West, 1992.

Popovic, Neil A. F. *Humanitarian Law, Protection of the Environment, and Human Rights.* 8 Georgetown International Environmental Law Review 67 (1995).

Renner, Michael. *Fighting for Survival: Environmental Decline, Social Conflict, and the New Age of Insecurity.* New York: W.W. Norton, 1996.

Roodman, David Malin. *The Natural Wealth of Nations: Harnessing the Market for the Environment.* New York: W.W. Norton, 1998.

Sachs, Aaron. *Eco Justice: Linking Human Rights and the Environment.* Washington, DC: Worldwatch Paper No. 127, 1995.

Sands, Phillipe. *Principles of Environmental Law I: Frameworks, Standards and Implementation.* Manchester, UK: Manchester University Press, 1995.

Schwabach, Aaron. *The Sandoz Spill: the Failure of International Law to Protect the Rhine from Pollution.* 16 Ecology Law Quarterly 443 (1989).

————. *Diverting the Danube: The Gabčíkovo-Nagymaros Dispute and International Freshwater Law.* 14 Berkeley Journal of International Law 290 (1996).

Schwebel, Stephen M. *The Effect of General Assembly Resolutions on Customary International Law,* Proceedings of the American Society of International Law 301 (1979).

Scott, Richard F., and Alain Levasseur. *Cases and Materials on the Law of the European Union.* St. Paul, MN: West, 2001.

Scott, Richard F., Edwin B. Firmage, Christopher L. Blakesley, and Sharon Williams. *The International Legal System, 5th ed.* New York: Foundation Press, 2001.

Sessions, George, ed. *Deep Ecology for the 21st Century: Readings on the Philosophy and Practice of the New Environmentalism.* Boston: Shambhala Publications, 1995.

Shahabuddeen, Mohamed. *Precedent in the World Court.* Cambridge, UK: Cambridge University Press, 1996.

Slaymaker, Olav, and Tom Spencer. *Physical Geography and Global Environmental Change.* Singapore: Addison Wesley Longman, 1998.

Slomanson, William R. *Fundamental Perspectives on International Law, 4th ed.* St. Paul, MN: West, 2002.

Stern, Paul C. et al., eds. *Environmentally Significant Consumption.* Washington, DC: National Academy Press, 1997.

Switzer, Jacqueline Vaughn. *Environmental Activism.* Santa Barbara: ABC-CLIO, 2003.

Teclaff, Ludwik A. *Fiat or Custom: The Checkered Development of International Water Law.* 31 Natural Resources Journal 45 (1991).

Teclaff, Ludwik A., and Eileen Teclaff. *Transboundary Toxic Pollution and the Drainage Basin Concept.* 25 Natural Resources Journal 589 (1985).

Tesón, Fernando R. *A Philosophy of International Law.* Boulder, CO: Westview Press, 1998.

Turner, Barry, ed. *The Statesman's Yearbook: The Politics, Cultures, and Economies of the World, 139th ed.* Houndmills, Basingstoke, UK: Palgrave, 2001.

Uhlman, Eva M. Kornicker. *State Community Interests, Jus Cogens and Protection of the Global Environment: Developing Criteria for Peremptory Norms.* 11 Georgetown International Law Review 101 (1998).

United Nations War Crimes Commission. *History of the United Nations War Crimes Commission and the Development of the Laws of War.* London: United Nations, 1948.

Weiss, Edith Brown, et al. *International Environmental Law and Policy.* New York: Aspen, 1998.

Williams, Sharon. *Public International Law Governing Transboundary Pollution.* 13 University of Queensland Law Journal 112 (1984).

Yuzon, Florencio J. *Deliberate Environmental Modification Through the Use of Chemical and Biological Weapons: "Greening" the International Laws of Armed Conflict to Establish an Environmentally Protective Regime.* 11 American University Journal of International Law and Policy 793 (1996).

Treaties and Other International Agreements

Act for the Navigation of the Danube between Austria, Bavaria, Turkey, and Wurttemberg, Nov. 7, 1857, art. XXIX, 117 Parry's T.S.471, 482 (French text).

Additional Protocol to the American Convention on Human Rights, Nov. 17, 1988, O.A.S. Treaty Series 69 (1988), 28 I.L.M. 156.

African Charter on Human and Peoples' Rights. OAU Doc. CAB/LEG/67/3/CAB/LEG/67/3 rev. 5, 21 I.L.M. 59 (1982).

Armistice Convention With Austria-Hungary, Naval Conditions, Nov. 3, 1918, United States Senate: Treaties, Conventions, International Acts, Protocols, & Agreements 3529.

Arrangement and Final Protocol Relative to the Exercise of the Powers of the European Commission of the Danube, Aug. 18, 1938, art. 12, 196 L.N.T.S. 113 (Treaty of Sinaia).

Articles Between Austria, Bavaria, Turkey, and Wurttemberg Additional to the Act for the Navigation of the Danube [of] Nov. 7, 1857, Mar. 1, 1859, art. III, 120 Parry's T.S. 275, 277 (French text).

Articles of Agreement of the International Monetary Fund, July 22, 1944, 60 Stat. 1401, 2 U.N.T.S. 39, amended by Amendment of the Articles of Agreement of the International Monetary Fund, May 31, 1968, 20 U.S.T. 2775, 726 U.N.T.S. 266; Second Amendment to the Articles of Agreement of the International Monetary Fund, Apr. 30, 1976, 29 U.S.T. 2203, T.I.A.S. No. 8937; Third Amendment to the Articles of Agreement of the International Monetary Fund, Nov. 11, 1992, 31 I.L.M. 1307.

Cartagena Protocol on Biosafety to the Convention on Biological Diversity, Jan. 29, 2000, 39 I.L.M. 1027.

Convention Concerning Fishing in the Waters of the Danube, Jan. 29, 1958, 339 U.N.T.S. 58.

Convention Concerning the Protection of the World Cultural and Natural Heritage, Nov. 16, 1972, 27 U.S.T. 37, 1037 U.N.T.S. 151.

Convention Instituting the Definitive Statute of the Danube, July 23, 1921, 25 L.N.T.S. 173.

Convention on the Rights of the Child, Nov. 20, 1989, 1577 U.N.T.S. 3.

Convention Regarding the Regime of Navigation on the Danube, Aug. 18, 1948, 32 U.N.T.S. 181 (English text begins at page 197), 33 U.N.T.S. 201.

Espoo Convention on Environmental Impact Assessment in a Transboundary Context, Feb. 25, 1991, 30 I.L.M. 800 (1991).

Military Convention Under Which the Armistice Signed Between the Allies and Austria-Hungary Is To Be Applied In Hungary, Nov. 13, 1918, United States Senate: Treaties, Conventions, International Acts, Protocols, & Agreements 3537.

Montreal Protocol on Substances that Deplete the Ozone Layer, Sep. 16, 1987, 26 I.L.M. 1550 (1987).

Public Act of the European Commission of the Danube Relative to the Navigation of the Mouths of the Danube, Nov. 2, 1865, 131 Parry's T. S. 399, 422–23.

Regulations of Navigation and Police Applicable to the Danube Between Galatz and the Mouths, Drawn up by the European Commission of the Danube, May 19, 1881, arts. 26, 73, 158 Parry's T.S. 245.

Resolution on the Institutional and Financial Arrangements for International Environment Cooperation (establishing the United Nations Environment Program), Dec. 15, 1972, G.A. Res. 2997, UN GAOR, 27th Sess., Supp. 30, at 42, UN Doc A/8370 (1973), 13 I.L.M. 234 (1974).

Rome Statute on the International Criminal Court, UN Doc. A/CONF. 183/9 (1998).

Statute of the International Court of Justice, 1976 Y.B.U.N. 1052, 59 Stat. 1031, T.S. No. 993.

Stockholm Convention on Implementing International Action on Certain Persistent Organic Pollutants, May 22, 2001, 40 I.L.M. 532 (2001).

Treaty of Peace Between the Allied and Associated Powers and Hungary, June 4, 1920, United States Senate: Treaties, Conventions, International Acts, Protocols, & Agreements 3539 (Treaty of Trianon).

Treaty of Peace Between Austria-Hungary, Bulgaria, Germany, Turkey, and Romania, May 7, 1918, 223 Parry's T.S. 256.

Treaty of Peace With Germany, June 28, 1919, 225 Parry's T.S. 189 (Treaty of Versailles).

Treaty of Peace with Hungary, Feb. 10, 1947, 41 U.N.T.S. 135 (English text begins on page 168).

United Nations Convention on the Law of the Non-navigational Uses of International Watercourses. G.A. Res. 51/229, U.N. GAOR, 51st Sess., May 21, 1997; 36 I.L.M. 700 (1997).

United Nations Economic Commission for Europe Convention on the Protection and Use of Transboundary Watercourses and International Lakes, Mar. 17, 1992, 31 I.L.M. 1312.

United Nations Framework Convention on Climate Change, May 9, 1992, 31 I.L.M. 849 (1992).

Vienna Convention for the Protection of the Ozone Layer, Mar. 22, 1985, UNEP Doc. IG.53/5, 26 I.L.M. 1529 (1987).

World Trade Organization, Agreement on the Application of Sanitary and Phytosanitary Measures, Apr. 15, 1994, 1867 U.N.T.S. 493.

Other International Materials

Affaire du Lac Lanoux (Spain v. Fr.), 12 Reports of International Arbitral Awards 281 (1957), digested in 53 Am. Journal of International Law 56 (1959).

Agenda 21, June 13, 1992, U.N. Doc. A/CONF.151/26(1992).

Case Concerning the Jurisdiction of the International Commission of the River Oder, 1929 P.C.I.J. (Ser. A) No. 23, at 27 (Sept. 10, 1929).

Commission on Sustainable Development, *Overall Progress Achieved Since the United Nations Conference on Environment and Development*, E/CN.17/1997/2/Add. 17, Jan. 17, 1997.

Corfu Channel Case (U.K. v. Alb.), 1949 I.C.J. 4, 22 (1949) (determination on the merits).

Donauversinkung Decision (Baden v. Wurttemberg), 116 Entscheidungen des Reichsgerichts in Zivilsachen, Suppl. Entscheidungen des Staatsgerichtshofs 18; see also Annual Digest & Reporter of Public International Law Cases 128 (RGst. 1927). Discussed in detail in Johan G. Lammers, *Pollution of International Watercourses* 433–36 (Dordrecht, Netherlands: Martinus Nijhoff, 1984)

Draft Treaty Establishing a Constitution for Europe, July 18, 2003, CONV 850/03. Available at http://european-convention.eu.int/DraftTreaty. asp?lang=EN (visited May 30, 2004).

Experts Group on Environmental Law of the World Commission on Environment and Development, Legal Principles for Environmental Protection and Sustainable Development, Aug. 4, 1987, UN Doc. WCED/86/23/Add. 1 (1986).

International Monetary Fund. *The IMF and the Environment: A Factsheet, April 2004.* Available at http://www.imf.org/external/np/exr/facts/ enviro.htm (visited Sept. 10, 2004).

International Regulation Regarding the Use of International Watercourses for Purposes other than Navigation, 10 April 1911, Madrid, 24 Annuaire de l'Institut de droit international 365 (1911), reprinted in FAO Legislative Study No. 65 269 (1998).

Johannesburg Plan of Implementation, Sep. 4, 2002, U.N. Doc. A/CONF. 199/20.

Judgment of Feb. 13, 1939, Corte cass., Italy, 64 Foro It. I 1036, 1046, digested in 3 Digest of International Law 1050–51 (1938–39).

Jurisdiction of the European Commission of the Danube Between Galatz and Braila, Advisory Opinion, 1927 P.C.I.J. (ser. B) No. 14.

Legality of the Threat or Use of Nuclear Weapons (advisory opinion), 35 I.L.M. 809, 824 (July 8, 1996).

Making Sustainable Commitments: An Environment Strategy for the World Bank. Available at http://www-wds.worldbank.org/servlet/WDS_ IBank_Servlet?pcont=details&eid=000094946_01110704111523 (visited May 28, 2004), June 19, 2001, available at http://www.pca-cpa.org/EN-GLISH/EDR/ (visited May 28, 2004).

Report of the United Nations Stockholm Conference on the Human Environment, U.N. Doc. A/CONF.48/14/Rev. 1 (1973), 11 I.L.M. 1416 (1972) [Stockholm Declaration].

Rio Declaration on Environment and Development, June 14, 1992, U.N. Doc. A/CONF.151/26 (vol. I), 31 I.L.M. 874 (1992).

Texaco Overseas Petroleum Co. v. Libyan Arab Republic, 17 I.L.M. 1, 27 (1978).

Trail Smelter Case (U.S. v. Can.), 3 R.I.A.A. 1905, 1965 (1941). Reprinted in 35 American Journal of International Law 684 (1941).

UNEP Ozone Secretariat, Press Release: Update on this Year's Antarctic Ozone Hole on the Occasion of the International Day for the Preservation of the Ozone Layer, Sept. 15, 2000. Available at http://www.unep.org/ozone/Events/WMOpressrel_ozoneday2000.asp (visited Sept. 12, 2004).

World Charter for Nature, Oct. 28, 1982, G.A. Res. 37/7 (Annex), U.N. GAOR, 37th Sess., Supp. No. 51, at 17, U.N. Doc. A/37/7, 22 I.L.M. 455 (1983).

2

Problems, Controversies, and Solutions

Problems and controversies exist in every area of international environmental law; solutions, unfortunately, are not as common. Most international environmental disputes are either local—that is, confined to an area along or near a national border —or are expressions of the underlying tension between environment and development; often they are both. This chapter looks first at an example of this latter sort of problem: a dispute between two neighboring countries, Hungary and Slovakia, over a planned hydroelectric project crossing their shared border, with the involvement of a wealthy neighbor, Austria, seeking to export some of its own environmental difficulties. Because relations between the neighbors were peaceful, and both parties had a strong commitment to the international rule of law, the dispute was submitted to the International Court of Justice (ICJ). The ICJ's decision clearly stated the rights and responsibilities of the parties under international law but does not seem to have solved the problem. The disastrous environmental legacy of Communism persists in Central and Eastern Europe, hampering economic development.

This chapter also looks at international trade and its relation to economic development and thus to the environment, focusing on sub-Saharan Africa. In much of the South, especially in Africa, continuing poverty is the result of barriers to market access. If developing African nations had greater access to northern markets for their agricultural produce, they would benefit economically. The failure of northern countries to open their markets reflects the lobbying power of northern agricultural interests more than any misguided environmentalist desire to reduce consumption by limiting wealth, but it fuels widespread resentment in the South based on the belief that the North wants the South to remain poor. In this case any initiative for change must come from the North, and at present at least the problem seems to be beyond solution.

The next geographic area examined, however, presents a success story: Antarctica. In Antarctica a territorial dispute with environmental consequences has been resolved by the creation of a treaty regime firmly committed to environmental protection. And attempts to protect the atmosphere from pollution have had mixed results: Success in addressing chemical and particulate pollution and ozone depletion, but (so far) failure to address global warming.

Attempts to address the other problems discussed in this chapter have met with mixed success and have sometimes led to new problems. Well-intentioned efforts to protect biodiversity inadvertently gave rise to a conflict between the developed North and the developing South over bioprospecting and biopiracy. So far the object of the conflict has not turned out to be as valuable as both sides initially thought; a great deal of effort was expended, and good will squandered, by all involved, and little was gained.

Freshwater and ocean environmental law, discussed in other volumes in the Contemporary World Issues Series, are briefly examined, as are two successful attempts by countries of the South to challenge regimes favorable to the North. In one case, a group of developing countries in Africa banded together to resist a waste-disposal regime that might have resulted in their countries becoming toxic waste sites. In the other, the tiny Pacific island nation of Tonga successfully asserted its interests in resources located in outer space, in defiance of a regime favoring more developed countries.

One area of international environmental law discussed in Chapter 1 is not discussed here: War is a particular problem for the United States, the world's greatest military power, and the United States has found itself embroiled in a complex international environmental dispute resulting from the destruction of a chemical complex in the former Yugoslavia during the Kosovo War. This conflict is discussed in detail in Chapter 3.

The Classic Model of International Dispute Resolution: The Gabčíkovo-Nagymaros Dispute

The conflict between development and environmental protection is especially sharply delineated in the former Communist countries of Europe. During the Communist era, extending human control over natural systems was viewed as an end in itself; governments of the time favored massive industrial projects that often produced little or no economic benefit, and caused untold environmental damage. Toward the end of the Communist era, the environmentalist movement became closely identified with resistance to the regime in some countries, perhaps most notably in Hungary.

The post-Communist era has left these countries with a legacy of environmental disaster. Some, such as Poland and Hungary, are now members of the North Atlantic Treaty Organization (NATO) and the European Union (EU)—full-fledged members of the "North," with northern environmental concerns; the problem for these countries is how to obtain sufficient funding to clean up the mess they have inherited. Others, such as Romania, find that exploitation and mismanagement by the previous regime have left them in a state of intense poverty—they are now Third World nations, part of the global "South," with environmental problems as severe as those of Poland or Hungary, but with insufficient resources to address those problems and with southern development concerns taking precedence even over environmental cleanup and certainly over environmental preservation.

The case of the Gabčíkovo-Nagymaros project is important both because it provides a look at the environmental problems of the region before and after the collapse of Communism and because it provides a rare example of an environmental dispute that has gone through all of the stages of the most formal process of dispute resolution available under the auspices of the United Nations: After other dispute-resolution mechanisms, including negotiation and a trilateral investigating committee, with the European Community as the third party, had failed, the parties litigated the matter before the ICJ, which then issued a decision on the merits.

The Danube River is one of Europe's great waterways; for millennia it has carried passengers and freight between the countries along its banks from Germany to the Black Sea. But between Gabčíkovo, in what is now Slovakia, and Nagymaros, Hungary, the Danube was historically only intermittently navigable, especially in the region known as the Szigetköz. In the Szigetköz, the river spreads out into numerous strands threaded with islands. The Szigetköz was one of the world's largest inland deltas and one of Central Europe's few remaining large wetlands; at one time it provided a home to more than 5,000 species of animals. But during much of the year it was also an obstacle to navigation. In 1951, the governments of Hungary and what was then Czechoslovakia began planning to eliminate this obstacle. At the time, both countries were governed by Communist governments more or less under the control of the Soviet Union; both would rebel, unsuccessfully, against Soviet authority, Hungary in 1956 and Czechoslovakia in 1968 (Fitzmaurice, 1996).

The Gabčíkovo-Nagymaros project reflected a common theme in both communist and capitalist ideology in the first two-thirds of the twentieth century: The transformation of natural systems into systems under human control was seen not as an undesirable consequence of development but as something desirable, as Progress, and, especially in the Communist countries, often as an end in itself without regard to the profitability of the project.

So the Gabčíkovo-Nagymaros project was designed to be big, in the tradition of Communist hydraulic engineering projects such as the Iron Gates project in Romania and what was then Yugoslavia. Little happened immediately, however. In 1963 the two countries began negotiating a plan that would allow for the diversion of the Danube's waters without unduly infringing upon the sovereignty of either, and by 1968 (on the eve of the Soviet invasion of Czechoslovakia) the two countries had established a special River Administration for the sector of the Danube that included the area to be affected by the proposed Gabčíkovo-Nagymaros project.

It was the oil crisis of the 1970s that truly set the project in motion. The emphasis of the project shifted from navigation to energy production; plans were agreed upon in 1976 and formalized in a treaty between the two countries in 1977. Under the 1977 treaty, the Gabčíkovo-Nagymaros project would divert the waters of the Danube through a series of dams and reservoirs extending over 124 miles of the Danube valley between Bratislava (now the capital of Slovakia) and Nagymaros. The bulk of the construction would take place in Czechoslovakia, beginning with a 23-square-mile reservoir just outside of Bratislava. Most of the water collected in this reservoir would then be diverted into the diversion canal, a gigantic concrete bathtub 15.5 miles long: an aboveground aqueduct ranging in width from 900 to 2,000 feet and with walls up to 60 feet high, lined with plastic and carrying within it more than 90 percent of the natural flow of the Danube. Only 2.5 percent to 10 percent of the flow would be returned to the original riverbed, leading inevitably to the drying out of the Szigetköz. The diversion canal would solve the original problem—the obstacle to navigation provided by the Szigetköz (Heywood and Ravasz, 1989).

The second, and now more important, goal of the project—power generation—would be fulfilled by the hydroelectric plant to be built at Nagymaros, in Hungary. The 160-megawatt Nagymaros plant was designed to operate, not with a steady flow of water, but with periodic surges. Water would be released from the reservoir south of Bratislava to flow downstream through the diversion canal and 68 miles of riverbed as a flood wave to Nagymaros. An additional dam at Nagymaros would raise the water level in the riverbed between the diversion canal and Nagymaros, enhancing navigability (Heywood and Ravasz, 1989).

The Czechoslovakian government began construction at Gabčíkovo in 1978. Under the treaty, Hungary was to perform all of the work on its side of the border and 12 percent of the work on the Czechoslovakian side of the border; by 1981, three years later, the Hungarian government had barely begun construction.

Both countries at the time were still governed by Communist regimes, but Hungary's perhaps allowed more scope for expression than did Czechoslovakia's. At a public debate in 1980, many Hungarian scientists and engineers had openly criticized the project, and in November 1981, journalist Janos Vargha published an article in the Hungarian magazine *Valosag* ("Truth") condemning the project; the article was to do for Hungary what Rachel Carson's *Silent Spring* had done for the United States, and more.

Vargha's article helped to launch a broad-based environmental movement in Hungary, with the Nagymaros project as its first and central target; the Hungarian government stopped construction on Nagymaros in 1981. In 1982, a new Hungarian commissioner was appointed, ostensibly to coordinate the project with Czechoslovakia, although he later stated that his understanding was that his task was "to stop the project or put it off." In the meantime, however, Czechoslovakia had begun construction and had performed some work initially allotted to Hungary; by 1983 it had persuaded Hungary to resume work. In February 1984, Janos Vargha and others founded Soviet-bloc Europe's first grass-roots environmental organization, which eventually became the Duna Kör (Danube Circle) (Galambos, 1992). The protests led by Duna Kör would eventually be instrumental in the collapse of the Communist regime in Hungary.

Meanwhile, another country was taking an interest in Gabčíkovo-Nagymaros. Neighboring Austria, a wealthy democratic state, had already exploited most of its hydropower resources; environmentalists strongly opposed further development within Austria. In the mid-1980s protesters succeeded in stopping the construction of a hydroelectric plant in an Austrian nature reserve; subsequently, in May 1986, Austria and Hungary agreed that Austrian banks would loan money for the building of the plant at Nagymaros. Seventy percent of all building contracts would go to Austrian companies, and Hungary would supply Austria with 1.2 billion kilowatt hours per year of electricity (Schapiro, 1990). This was one-third of the total expected output of the Gabčíkovo-Nagymaros project; it was quite possible that Hungary would have to import electricity to meet its obligations to Austria.

Austria's involvement in the Gabčíkovo-Nagymaros project thus represented another aspect of the North-South conflict; not the developed North seeking to oppose development in the South, but the developed North exporting environmentally unsound industries to the South. The people of Austria were unwilling to accept another hydroelectric plant within their own borders but still wanted electricity; in neighboring Hungary, poorer than Austria but perhaps most importantly with a reputation for suppressing political dissent, the opposition of the people might be unable to halt the power plant.

The political climate had changed dramatically since the beginning of the 1980s, however. When Austrian contractors sought to resume work at Gabčíkovo, they were immediately met by organized opposition from Duna Kör and other environmental groups. The project had seemed profitable to Austria in part because of the supposed absence of protesters under an authoritarian regime; the contractors were taken aback at the strength of the opposition to the project. Although the Hungarian Parliament had voted as late as 1988 to continue the project, in March 1989 the prime minister announced that the government was willing to consider a referendum on Gabčíkovo-Nagymaros. The announcement was but one facet of a far greater change: the collapse of communism in Hungary, and across much of the world, in 1989. Immediately after the announcement four members of the government, all supporters of the Gabčíkovo-Nagymaros project, resigned; among these were the speaker of the parliament and a former head of state. Two months later the Hungarian government stopped work on the project completely. Throughout that summer and fall the two countries continued to wrangle over the project, while simultaneously their systems of government were changing. In October, the Hungarian constitution was amended to allow for free elections, and the parliament voted overwhelmingly to withdraw from the Gabčíkovo-Nagymaros project (Fitzmaurice, 1998).

This left the Czechoslovakian government with a nearly completed water diversion project and Gabčíkovo plant, and no power plant at Nagymaros. Rather than abandon the project, Czechoslovakia designed a "provisional solution": building additional structures in Czechoslovakia that would make it possible to run the Gabčíkovo plant, albeit less efficiently, without Nagymaros. In December, Czechoslovakia's "Velvet Revolution" took place, and Czechoslovakia also abandoned one-party rule.

In Hungary, opposition to the Gabčíkovo-Nagymaros project had been nearly synonymous with opposition to the old regime. Many Hungarians therefore expected that Czechoslovakia would feel similarly and, having rejected Communism, would also reject the dam. They were half right: There was little support for the project among the Czechs. In Slovakia, however, Gabčíkovo became a symbol of Slovak nationalism, closely identified with Slovakia's efforts to achieve independence and establish itself as a nation. It was also an essential resource for a region that, if established as an independent country, would otherwise suffer from a severe shortage of electricity. Despite ongoing protests by Czechoslovakian and Hungarian environmentalists, work at Gabčíkovo continued through 1990 and 1991. To discourage Slovak separatism, the Czechoslovakian government continued to support the project. In November of 1991, Czechoslovakia began construction of the additional structures required by the "provisional solution" (Liska, 1992). In May 1992, Hungary informed Czechoslovakia that it was unilaterally terminating the treaty regarding construction of the Gabčíkovo-Nagymaros project, and in July 1992, Slovakia declared its "sovereignty," and the two parts of Czechoslovakia agreed on a plan for partition. In October, Slovakia began diverting water from the Danube, and on 1 January 1993, Czechoslovakia's "Velvet Divorce" became final; Slovakia was an independent nation, and the concerns of the new Czech Republic became irrelevant to the future of Gabčíkovo.

Slovakia's "provisional solution" was an environmental disaster. Among the environmental problems it caused were the destruction of wetlands; contamination and significant loss of the Szigetköz aquifer; fluctuation of the water level of the Danube and contamination of the river banks by sewage, especially at Gyor, Hungary, and contamination of other water sources as far downstream as Budapest; the drying out of thousands of hectares of forest and productive farmland; and a general decline in the flora and fauna of the region.

The two states that were parties to the dispute, Hungary and Slovakia, agreed to submit the dispute to the ICJ by special agreement, one of the three ways in which states can consent to the ICJ's jurisdiction. In the special agreement, the parties asked the ICJ to decide three questions: whether Hungary was entitled to abandon the Nagymaros portion of the Gabčíkovo-Nagymaros project; whether Czechoslovakia, and later Slovakia, was entitled to implement its "provisional solution"; and what the legal effects of Hungary's 1992 notice of termination were.

The ICJ, after hearing argument and receiving submissions from both parties and conducting an unprecedented visit to the site of the project, held that Hungary was not entitled to suspend and later abandon work on its portion of the Gabčíkovo-Nagymaros project. The Court rejected Hungary's argument that by abandoning work on the project it did not reject the 1977 treaty, as the "effect of Hungary's conduct was to render impossible the accomplishment of the system of works that the Treaty expressly described as 'single and indivisible'." It then turned to the question of whether Hungary's *de facto* rejection of the treaty was justified by necessity.

To be justified by necessity, the ICJ stated, a state's action must be the only means of safeguarding an essential state interest threatened by a grave and imminent peril to which the state's actions must not have been a contributing cause; and the act complained of must not seriously impair an essential interest of another state. The ICJ found that "the concerns expressed by Hungary for its natural environment in the region affected by the Gabčíkovo-Nagymaros Project related to an 'essential interest' of that State," but it also found that Hungary's actions had contributed to the peril and that they might seriously impair Slovakia's interests. Environmentalists were particularly distressed by the ICJ's determination that in 1989 the gravity of the perils had not been sufficiently established, nor was the harm imminent; to many this seemed to show a lack of regard for environmental concerns.

But the ICJ also concluded that Slovakia (as well as Czechoslovakia before it) violated its obligations under the 1977 treaty, and thus international law, by putting the "provisional solution" into effect. The ICJ pointed out that the treaty provided "for the construction of the Gabčíkovo-Nagymaros System of Locks as a . . . single and indivisible operational system of works" that "could not be carried out by unilateral action." In other words, Hungary had been wrong to abandon work at Nagymaros, but Slovakia had also been wrong in proceeding with its provisional solution.

On the third issue, Hungary had presented five arguments for the legality and validity of its 1992 notice of termination on the 1977 treaty: necessity, impossibility of performance, fundamental change of circumstances, material breach by Czechoslovakia, and the development of new norms of international environmental law. Necessity the Court had already addressed; it noted in addition that necessity is a defense that may be raised after a state has failed to comply with its obligations under a treaty but is not grounds for termination of a treaty.

The arguments relating to impossibility of performance and changed circumstances were only tangentially environmental issues; they referred to the collapse of Communism in Hungary (and across the world) at the end of the 1980s and beginning of the 1990s. These questions obviously had sweeping implications, though; the six (seven, after the breakup of Czechoslovakia) former Warsaw Pact nations, let alone the fifteen that sprang from the former Soviet Union, the five (with perhaps more to come) that sprang from the former Yugoslavia, and Albania and Mongolia had been parties to hundreds of treaties; the ICJ may have been influenced by the knowledge that a decision in favor of Hungary on these two issues would undercut all of those treaties, leaving those states and any others with which they had entered into international agreements prior to the collapse of Communism uncertain as to what obligations existed in a wide range of fields.

The ICJ found that the material breach by (Czecho)slovakia consisted not of notifying Hungary of its intention to implement the provisional solution, but of actually doing so. This happened in October 1992, five months after Hungary's notice of termination, and thus could not have entitled Hungary to terminate the treaty prior to that time.

Another disappointment for environmentalists was the ICJ's treatment of the "new norms of international environmental law" argument. Although, as we saw in the previous chapter, Judge Weeramantry stated in a separate opinion that sustainable development is in fact a principle of customary international law, the majority of the ICJ decided otherwise. Even if the ICJ had found that such a principle existed in international custom, it would not have allowed Hungary to avoid its obligations under the 1977 treaty: The ICJ took the position that while certain peremptory norms (also known as *jus cogens* norms) of international law might void a treaty, no such norm had emerged in this case. The 1977 treaty itself provided a mechanism by which the parties could take into account other changes in the norms of customary international law, and thus Hungary was not justified in attempting to terminate the treaty as a result of any such changes. The ICJ also rejected the idea that the "disappearance of one of the parties" had negated the treaty; "treaties of a territorial character have been regarded both in traditional doctrine and in modern opinion as unaffected by a succession of States," and the Gabčíkovo-Nagymaros treaty was of an evident territorial character.

The ICJ found that, although both parties had violated the 1977 treaty, "this reciprocal wrongful conduct did not bring the Treaty to an end nor justify its termination." The treaty established a regime for the management of a portion of the Danube. Slovakia's operation of the provisional solution was a continuing violation of the treaty and should be altered to conform to the treaty: "Re-establishment of the joint régime will also reflect in an optimal way the concept of common utilization of shared water resources for the achievement of the several objectives mentioned in the Treaty."

The ICJ recognized that environmental issues were central to the dispute but believed that they could be resolved within the bounds of the Gabčíkovo-Nagymaros treaty: "In order to evaluate the environmental risks, current standards must be taken into consideration. This is not only allowed by the wording of Articles 15 and 19, but even prescribed, to the extent that these articles impose a continuing—and thus necessarily evolving—obligation on the parties to maintain the quality of the water of the Danube and to protect nature." New norms and standards contained in instruments promulgated since 1977 would "have to be taken into consideration, and . . . given proper weight. . . . For the purposes of the present case, this means that the Parties together should look afresh at the effects on the environment of the operation of the Gabčíkovo power plant. In particular they must find a satisfactory solution for the volume of water to be released into the old bed of the Danube and into the side-arms on both sides of the river."

The essential thing was that the treaty must not be cast aside, but "that the Parties find an agreed solution within the co-operative context of the Treaty." Hungary and Slovakia were required to "negotiate in good faith in the light of the prevailing situation" and establish "a joint operational régime . . . in accordance with the Treaty[.]" Unless the parties were to agree otherwise, Hungary was to compensate Slovakia for the damage sustained as a result of Hungary's suspension and subsequent abandonment of work on the project, while Slovakia was to compensate Hungary for damage caused by the provisional solution. The ICJ did not address the question of what specific amount of compensation should be paid, and by whom, but observed that "the issue of compensation could satisfactorily be resolved in the framework of an overall settlement if each of the Parties were to renounce or cancel all financial claims and counter-claims. At the same time, the Court wishes to point out that the settlement of accounts for the construction of the works is different from the issue of compensation, and must be resolved in accordance with the 1977 Treaty and related instruments."

Despite the willingness of both parties to submit the dispute to the ICJ and the court's even-handedness, the ICJ's decision did not end the dispute. The parties failed to reach an agreement, and within a year Slovakia had filed a request for an additional judgment; the case remains pending. Meanwhile negotiations between the parties continue.

Other International Environmental Problems in Central and Eastern Europe: The Tisza Cyanide Spill

Central and Eastern Europe, perhaps more than any other large areas of the Earth, except for the Soviet Union itself, were environmentally devastated during the second half of the twentieth century. Soviet domination of the region produced an unusual combination: highly industrialized countries with undemocratic governments. The industrial economies of the region had enormous power to cause environmental damage, and the authoritarian governments provided few or no channels for those concerned about environmental damage to protest, oppose, or even share information about the damage. Compounding the problem was the Stalinist ideal, seen in the Gabčíkovo-Nagymaros project, of transformation of nature as an end in itself. On emerging from authoritarian rule in the 1990s, the new governments found that they had inherited some of the world's most badly contaminated hazardous waste sites, along with highly polluting factories, mines, and automobiles (Feshbach and Friendly, 1992). Some, especially the poorer countries of the South, found that they lacked the resources to address these problems and were in such dire economic straits that they had little recourse other than to keep environmental enforcement lax in an effort to lure in foreign investors. This has led to a continuing series of environmental disasters in Romania and the former Yugoslavia, perhaps outweighed by less dramatic but continuous harm from the daily operation of environmentally unsound facilities.

Among the most dramatic environmental catastrophes in the poorer countries of post-Soviet Eastern Europe has been the 2000 Baia Mare spill, which contaminated the Tisza and the Danube. The Tisza spill was one of the most severe of many similar incidents in Romania and the former Yugoslavia, but it focused world attention on the dangerous combination of an aging Stalinist industrial plant and a complete lack of money to improve environmental safety. Beginning on 31 January 2000, at least 100,000 cubic meters of highly polluted water escaped from a tailings dam at the Aurul gold mine in Baia Mare, Romania.

The water flowed into the Somes, Tisza, and Danube rivers, causing enormous environmental damage. Most of the damage occurred in Hungary, downstream from Baia Mare (Mann, 2000; Middleton and Kemp, 2000).

The Aurul mine was a joint venture between a Romanian state-owned mining company and Esmeralda Mining, an Australian company; the disaster not only raised questions about Romania's obligations to Hungary, but about Australia's obligations to Romania and Hungary as an investor. The law on the former point, as we have seen, is quite well established, and Romania violated it by allowing its territory to be used for acts harmful to Hungary. As a practical matter, Romania, one of Europe's poorest countries, was not well situated to compensate relatively wealthy Hungary. But Hungary's long-term concern was not compensation for one particular incident, but prevention of similar incidents in the future. A UNEP team inspected the Baia Mare site later that year, and the two countries strengthened the legal regime protecting their shared portions of the Danube and Tisza rivers with a protocol on the prevention of environmental pollution signed at Debrecen, Hungary.

On the other side of the globe, Australia found itself politically divided by the catastrophe. Australian businesses that were in the habit of investing in overseas industry were horrified at the thought of Australian government controls on the environmental aspects of their overseas activities; such extraterritorial application of Australian environmental laws would, they claimed, put Australian business at a disadvantage in competition with businesses from other countries facing no such restrictions from their own governments. Australia's Greens and Democrats, on the other hand, took the moral high ground and insisted that Australia's environmental laws should be applied to Australian mining companies operating overseas just as they would be to activities in Australia (Middleton and Kemp, 2000).

The Baia Mare spill presents the same problem for Australia as the Gabčíkovo-Nagymaros project did for Austria. Australians were engaging in activity in the global "South" (Romania, of course, lies in the geographic, if not political and economic, North) that the Australian people deemed too environmentally risky for Australia. The proposed remedy—extraterritorial application of environmental laws—is one that has also been proposed for the United States by U.S. environmentalists.

Agricultural Policy and Foreign Trade: Irreconcilable Differences?

Among the bases for sustainable development enunciated in the Rio Declaration is the end of poverty. Principle 5 declares that "[a]ll States and all people shall cooperate in the essential task of eradicating poverty as an indispensable requirement for sustainable development, in order to decrease the disparities in standards of living and better meet the needs of the majority of the people of the world."

At the same time, some environmentalists in the North fear (and some are so tactless as to state openly) that increased wealth will lead to increased consumption and thus increase the "load" humanity places on the environment, even without any increase in population. One North American suburbanite, the argument runs, places a far greater load on the environment than one Botswanan farmer; providing the Botswanan farmer with the same level of wealth as a North American suburbanite will be environmentally disastrous.

This, of course, is what has led many developing country leaders and citizens to reject the idea of sustainable development: a fear that the North wishes the South to remain poor. It may also be unfounded; increasing wealth also seems to lead to increasing environmental awareness, and many of the excesses of consumption in the United States are the result of poor regulation and management rather than of anything inherent in the nature of a wealthy society. And poverty, like wealth, produces its own environmental problems; just as the gas-guzzling SUVs so common in the United States are a product of wealth, problems like slash-and-burn agriculture or the contamination of water by untreated sewage are products of poverty.

The primary mechanism by which poor countries may become wealthy is international trade; understanding and exploitation of absolute and comparative advantage has been the way in which many poor countries, especially in East Asia and, to a lesser extent, in Latin America, have emerged or are emerging from poverty.

The conflict between the North and South, and between northern environmentalists and southern economic needs, is at its starkest in the debate over agricultural policy and foreign trade. Terms like "fair trade," "green trade," and "sustainable trade" often serve to obscure the issues more than to enlighten, but to the extent that they reflect a desire both to remove rich-country barriers to poor-country trade and to ensure that poor people receive a substantial share of the benefit from that trade, they are admirable.

Sub-Saharan Africa is a region particularly grievously affected by trade barriers. Most of the world's poorest countries lie in sub-Saharan Africa. Most of sub-Saharan Africa is poor (Turner, 2001). The vast majority of the inhabitants of these poor countries are engaged in agriculture. For most of them this means subsistence agriculture, not because they are unable to grow cash crops, but because there is little or no market for cash crops. The local economies are too poor to provide much income; roads and other infrastructure are in poor condition, making it difficult to send produce to more distant markets in the same nation. And international trade is not an option because of trade barriers.

The major developed markets of the world—the United States, Japan, and the European Union—together spend an incredible amount of money—well in excess of $300 billion each year—subsidizing domestic agricultural production. This is more than six times the amount of foreign aid those countries provide to developing countries worldwide. Most of the crops that are subsidized could easily be grown in Africa; in fact, most already are (Diaz-Bonilla, 2003).

The actual cost to Africa, and to such other potential agricultural exporters as the countries of the Caribbean, may be much higher, however. Subsidies provide leverage. For example, suppose a domestic rich-world producer can grow, harvest, and bring to market a unit of a crop for $100. Now suppose that an African producer of the same crop, with higher shipping costs but lower land and labor costs, can bring a unit of the same crop to the same market for $90. In order to provide a level playing field for its own farmers, the rich-country government does not need to provide a subsidy equal to the full cost of production, or anywhere near it. A unit subsidy of $10 will enable the domestic producer and the African producer to sell at the same cost and receive the same profit. A subsidy of $11 per unit will effectively cut the African producer out of the rich-world market altogether; he will make $0. As a result, he will lack both the means and the incentive to develop new crops; purchase equipment, fertilizer, or pesticides; clear and plant new land; or otherwise increase his productivity.

The International Food Policy Research Institute estimates that rich-world agricultural subsidies result in the displacement of about $40 billion a year of agricultural imports from poor countries; more than half of this displacement is a result of EU subsidies, and another third is the result of U.S. subsidies (Diaz-Bonilla, 2003). Subsidies have another pernicious impact, however; they result in overproduction in rich countries. As a result, the United States and the EU export staple crops to the developing world; small local farmers cannot compete with subsidized milk and wheat dumped into developing-country markets.

In addition, not all trade barriers are subsidies; some are outright prohibitions. The United States, for example, sets quotas on the amount of sugar that may be imported each year; above that limit, no importation is permitted, at any price. Countries around the world routinely adopt "environmental protection regulations" or "sanitary and phytosanitary measures" in bad faith, intending not to protect themselves against some imagined danger to the domestic environment or to the health of domestic people, animals, or plants but to protect their farmers against competition from foreign trade. It is difficult to quantify the impact of non-tariff trade barriers on developing-country farmers.

Nor is direct lost income the only, or even the most important, measure of the economic harm done by subsidies. There are multiplier effects: For example, the dumping of subsidized EU sugar in Swaziland has meant the loss of 16,000 jobs in the sugar industry itself and an additional 20,000 jobs in connected industries such as packaging and transport. This total loss of 36,000 jobs in a country with a population of just over 1 million is roughly equivalent to the loss of over 10 million jobs in the U.S. economy. Those 36,000 workers are no longer spending their paychecks on consumer goods and services, nor will they be able to invest in businesses of their own. Their employers will not be expanding their businesses; instead, they will be selling off assets and equipment at reduced prices, further depressing the economy. And many of the 36,000 laid-off employees will return to subsistence farming, taking themselves out of the cash economy and depriving other farmers of customers for their produce.

Suppose that, rather than dumping its own subsidized sugar in Swaziland, the EU were to stop subsidizing sugar altogether. Sugar grown in Europe would no longer be cheaper than sugar grown in Swaziland. Rather than contracting, the sugar industry would be expanding. The laid-off workers could get their jobs back. More would be hired. Sugar growers could buy new land and new equipment; the European appetite for sweeteners is more than sufficient to consume all of the sugar that Swaziland is capable of producing. Foreign exchange would flow into, rather than out of, Swaziland; it could use its euros to purchase manufactured goods unavailable locally, and to educate its young people for careers that would produce more added value than sugarcane production.

This would, of course, entail hardship for European sugar producers; their business would contract just as the business of the Swaziland sugar producers is now contracting. But that hardship would be offset, from the point of view of Europe as a whole, by the benefit to European sugar consumers. And the economy of the EU is better able to absorb this blow than is the economy of Swaziland. The loss of a five-dollar-a-day job can be a disaster for a Swazi cane-cutter and his family; a European farm worker will be able to find other work.

Why, then, do subsidies persist? Few inhabitants of rich-world countries are farmers; most are taxpayers, and all are consumers of agricultural products. Thus most pay twice for subsidies: once directly through taxes, and once indirectly through paying a higher price for domestic agricultural products than they would have paid for imports, in the absence of subsidies and trade barriers. Getting rid of subsidies would seem to be in the best economic interest of the majority of rich-country inhabitants.

The problem is that agricultural subsidies are politically entrenched. The people of the United States are represented according to population in the House of Representatives, but each of the fifty states, regardless of population, is represented in the Senate by two senators. States in which agriculture is the dominant business tend to have low populations; farms take up more land than factories or law firms. Thus agricultural interests are disproportionately represented in the Senate. Furthermore, a special-interest group that has enough members, is sufficiently well funded, and cares only about a single issue can generally manage to achieve its political will on that issue. Agriculture is bipartisan; farm interests have shown themselves equally friendly to Democrats and Republicans, as long as those Democrats and Republicans are in favor of price supports and trade barriers. Presidential elections create an added distortion: Because the presidential campaign season begins with the Iowa caucus and because Iowa is an agricultural state, each person who ultimately becomes president of the United States is forced to begin his campaign declaring his unwavering commitment to protecting the U.S. farmer.

There is also a sentimental attachment to the myth of the American family farm. Lobbyists for agricultural price supports conjure up images of the Joads in the *Grapes of Wrath*: "This is what free trade will do to the American family farmer!" (Admittedly the Joads are not the best example; their misfortunes were the result of environmentally unsustainable farming practices rather than of exclusion from markets.) The reality is that most agricultural production in rich countries is in the hands of large agribusiness concerns, and many "family farmers" are actually "weekend farmers"—wealthy individuals whose main interest in farming is as a tax write-off (another form of subsidy). Ironically, it is the poor farmers of the developing world who are most likely to be family farmers—often, like the Joads, forced off the land and doomed to wandering.

Europe and Japan have created similarly erroneous myths around their farmers. In addition, farming in Japan is linked in the popular imagination, or at least the political imagination, with national security. And the EU is a creature of treaty, dependent upon the consent of the states that make it up. The focus of the EU has been on eliminating internal trade barriers, rather than on eliminating trade barriers between the EU and the rest of the world. Barriers keeping non-European agriculture from the European market have often been politically necessary in order to gain the consent of agricultural states to the reduction or elimination of European trade barriers between EU members.

Among the most significant causes of the continuing poverty of much of Africa are war, corruption, and trade barriers. Only a third of these is under the control of northern governments. Some rich northern countries, such as the United States, have responded with trade preferences for manufactured goods from African countries. But countries with large populations of illiterate or poorly educated subsistence farmers are ill-suited to become the next Singapore; what their economies are currently best suited for is agriculture, and the North is unwilling to allow the agriculture of the South access to its markets.

Northern environmentalists do not reassure southern development ministers. It must be obvious to most southern farmers that exclusion from northern markets hurts them economically. It is also obvious to anyone who listens that northern environmentalists do not want Africans to farm more: If the people of Swaziland, Ethiopia, Zambia, Kenya, and Uganda, of all of sub-Saharan Africa, plant more sugar cane, corn, and wheat and raise more beef, where does that leave the elephant and the giraffe, the gazelle and the chimpanzee? To Africans it must seem that northern environmentalists who applaud "shoot-to-kill" orders given to game wardens must value the lives of lions and rhinoceroses above human life—at least, above African human life.

When asked to reconcile this conflict between environment and development and to propose a source of income for the inhabitants, the stock northern environmentalist response is "ecotourism." Apparently the entire population of Africa (or the entire rural population, which is most of the population) is supposed to transform itself into game wardens, tour guides, travel agents, hotel staff, waiters, and so forth. A deluge of northern tourists will then descend upon Africa, taking only pictures and leaving only footprints, and in the process showering the populace with dollars, euros, and yen. What works for St. Kitts or the Bahamas will work for sub-Saharan Africa, even if the former are islands of a few thousand people only an hour's flight from the North, while the latter is a vast, largely roadless continent that can take over 24 hours of travel to reach from the United States or Japan and that is inhabited by nearly a billion people. Apparently those northerners who want to spend days in transit for the chance to take a picture of a gorilla in its natural habitat are numerous enough and wealthy enough to support hundreds of millions of jobs in T-shirt shops and digital-photo printing.

International law has been almost entirely unsuccessful at addressing this problem. Of late, the World Trade Organization (WTO) has had difficulty even convincing its developed members not to raise trade barriers to imports of manufactured goods among themselves; too many of the developed members—nearly all, in fact—have politically entrenched but economically inefficient agricultural interests. And the traditional defenders of the interests of the poor and downtrodden have seized upon the idea that globalization of trade, and thus the WTO, is irredeemably evil and to be opposed at all costs. Although removing barriers to southern agricultural trade would solve many of the South's economic problems, it might well exacerbate some environmental problems, as environmentalists fear, even while improving the human environment and overall quality of life for many Africans. In any event, the political will to remove those barriers is nowhere to be found. Environmentalists and labor protectionists struggle against the WTO, confirming the belief of many in the South that not only the governments but also the people of the North wish the South to remain poor.

Antarctica: A Successful Mature Treaty Regime

Antarctica, like the high seas, lies outside the territory of any state. It is unique as the only continent, and the only large area of land, not to have been incorporated into some state's territory. Except for a few scientific research operations, it is uninhabited; as a practical matter, it is uninhabitable by humans. Its coasts, though, support an amazing quantity and variety of animal life.

Antarctica is the only continent to be completely governed under a treaty regime; Antarctica's "constitution" is the Antarctic Treaty, to which forty-five countries are signatories. These include most of the countries conducting research in Antarctica. At least one non-party that has conducted research in Antarctica— Pakistan—has recognized the authority of the treaty and complied with its terms in conducting its research.

The Antarctic Treaty regime is the result of the successful resolution of an international dispute: not an environmental dispute, but a more traditional territorial one. The original 1959 Antarctic Treaty was adopted in part to address the problem of territorial sovereignty over Antarctica. Seven states—Argentina, Australia, Chile, France, Norway, and the United Kingdom— claimed territory in the Antarctic; some of these claims overlapped. Five states refused to recognize any claims to Antarctica: Belgium, Japan, South Africa, the Soviet Union, and the United States. The treaty served in part to hold the claims of the seven claimant states in abeyance. This situation now seems likely to continue indefinitely (Guruswamy, 1997).

The original Antarctic Treaty was not truly an environmental treaty, although it did contain provisions prohibiting nuclear explosions and the disposal of radioactive wastes in Antarctica. The treaty identified "preservation and conservation of living resources" as an area of concern to the parties. At the third Antarctic Treaty Consultative Meeting the parties did in fact agree to Measures for the Conservation of Antarctic Fauna and Flora, designating Antarctica a "Special Conservation Area." And the later treaties that make up the present Antarctic Treaty regime are all environmental in nature: the Convention for the Conservation of Antarctic Seals, the Convention on the Conservation of Antarctic Marine Living Resources, and the Protocol on Environmental Protection to the Antarctic Treaty.

The Antarctic Treaty System is uniquely comprehensive in its rule-making, enforcement, and dispute-resolution powers. With an actual territory to govern, and without having to tread carefully around the sovereignty of national governments over that territory, the Antarctic Treaty System acts as something between a national government and a park service.

The Antarctic Treaty System is flexible, with no central administrative body; the equivalent of a Conference of Parties is the biennial Antarctic Treaty Consultative Meeting (ATCM). The treaty system includes the Antarctic Treaty itself and the associated treaties and protocols; recommendations introduced at the ATCMs, if subsequently ratified by all member states; decisions of Antarctic Treaty Special Consultative Meetings (ATSCMs); the results of Meetings of Experts; and information provided by bodies such as the Scientific Committee on Antarctic Research.

The Scientific Committee on Antarctic Research coordinates Antarctic research programs and encourages scientific cooperation; the Council of Managers of National Antarctic Programs, composed of the heads of each of the national Antarctic operating agencies, meets annually to exchange logistic information, encourage cooperation, and develop advice to the Treaty parties. Intergovernmental and nongovernmental organizations, including the International Union for the Conservation of Nature, the United Nations Environment Program, and the Antarctic and Southern Ocean Coalition, are invited to the Treaty meetings as experts, as are the International Hydrographic Organization, the World Meteorological Organization, the Intergovernmental Oceanographic Commission, and the International Association of Antarctic Tour Operators (Guruswamy, 1997).

The Antarctic Treaty regime is not a perfect solution; only those nations that conduct substantial scientific research (a difficult and expensive undertaking) in the Antarctic can be "consultative parties" with full voting rights. It is thus sometimes criticized as a rich and/or powerful countries' club, although the consultative parties include countries such as Bulgaria and Ecuador that hardly fit either description.

Atmospheric Pollution: Two Qualified Successes and One Partial Failure

The regime of customary international law that originated with the *Trail Smelter* dealt with regional pollution; it eventually matured into a body of regional law, largely codified in conventions, dealing with such problems as acid rain. In recent decades, however, concern over regional chemical and particulate pollution of the air has been eclipsed by two more dramatic concerns: depletion of the Earth's ozone layer and climate change caused by greenhouse gases. Each of these problems, if not checked, has the potential not only to drastically affect human health and economic activities, but also to render the Earth uninhabitable for the human race.

Transboundary Air Pollution

Regional conventions dealing with transboundary chemical and particulate air pollution are too numerous to list; customary international law on the topic, of course, dates back to the *Trail Smelter* arbitration. The closest thing to a global treaty is the Convention on Long-range Transboundary Air Pollution (LRTAP), adopted in 1979 and in force since 1983, to which forty-nine northern hemisphere countries are parties. The United States is also a party to a 1991 Agreement on Air Quality with Canada, which addresses the problem of acid rain, and a broader agreement, the 1983 Agreement to Cooperate in the Solution of Environmental Problems in the Border Area, with Mexico. (The latter is also known as the La Paz Agreement.) Two annexes to the La Paz Agreement specifically address air pollution; one addresses sulfur dioxide emissions from smelting plants (a cause of acid rain), and the other addresses urban air pollution.

The LRTAP is a fairly early post-Stockholm framework convention: It does not try to solve the problem with a one-time solution that may be unable to win widespread acceptance and does not take into account changes in scientific understanding, consumption and production patterns, or industrial technology. Instead, the framework convention was viewed as a starting point, and it was anticipated that the contracting parties would negotiate additional protocols as necessary. This they have done: The parties to the convention have managed, through eight protocols, to establish emission standards for a wide range of air pollutants. The eight protocols aim to reduce emissions of sulfur, nitrogen oxides, volatile organic compounds, heavy metals, and persistent organic pollutants, as well as addressing acidification, eutrophication, and ground-level ozone (which, in contrast to stratospheric ozone, is harmful) and financing monitoring and evaluation efforts.

The LRTAP and its eight protocols have created several intergovernmental bodies, expert groups, and scientific centers. In the event of noncompliance with the convention, the LRTAP somewhat nebulously provides that the executive body may "decide upon measures of a non-discriminatory nature to bring about full compliance with the protocol in question, including measures to assist a Party's compliance." The LRTAP's dispute-resolution provision is equally perfunctory: "If a dispute arises between two or more Contracting Parties to the present Convention as to the interpretation or application of the Convention, they shall seek a solution by negotiation or by any other method of dispute settlement acceptable to the parties to the dispute."

Ozone Depletion

As with transboundary air pollution, international environmental law's efforts to address ozone depletion have met with a fair degree of success. Ozone molecules in the earth's stratosphere—the ozone layer—are necessary to protect humanity and probably most diurnal land animals. Harmful, even lethal, levels of ultraviolet radiation reach the Earth's upper atmosphere from the sun. The ozone layer filters this radiation so that the amount reaching the surface is relatively harmless. A decrease in the amount of stratospheric ozone means an increase in the amount of ultraviolet radiation reaching the earth. More ultraviolet radiation means more skin cancers, cataracts, and immune system dysfunction for humans and other animals, and harm to some plants on which animals and humans depend for food.

Ozone (O_3) is constantly being formed in the upper atmosphere by the action of solar ultraviolet radiation on ordinary molecules of oxygen (O_2). The ozone molecule is inherently unstable and can easily be broken down by interaction with a variety of naturally occurring gases. The dynamic equilibrium thus created results in a stratospheric ozone concentration that fluctuates seasonally but should be relatively constant over the long term, absent the introduction of new gases that may increase the rate of ozone destruction.

Anthropogenic chlorine compounds, particularly chlorofluorocarbons (CFCs), have accelerated the rate of ozone destruction. However, CFCs and related chemicals have important economic uses, especially in refrigeration; one of the tasks of international law is to provide a mechanism for adequately balancing the potential harm from ozone depletion against the known benefits of ozone-depleting chemicals, taking into account the cost of developing, producing, and using substitutes (D'Amato and Engel, 1996).

Ozone depletion is the subject of the 1985 Vienna Ozone Convention and the convention's 1987 Montreal Protocol on Substances that Deplete the Ozone Layer, to which 184 countries have become parties. The original Vienna Ozone Convention was adopted before the harmful effects of ozone-depleting chemicals were actually known; it is thus an example—perhaps the best example—of application of the precautionary principle. The Vienna Ozone Convention contains a vague agreement among the parties to take "appropriate measures" for the protection of "human health and the environment against adverse effects resulting or likely to result from human activities which modify or are likely to modify the Ozone Layer[.]" Its primary purpose, however, was to encourage research and monitoring and to set the stage for further cooperation. This took the form of the 1987 Montreal Protocol, which has been adjusted or amended five times since its adoption.

The Montreal Protocol provides measures for the control of nine categories of ozone-depleting chemicals, currently containing ninety-six chemicals in all, with the ultimate goal of completely eliminating these chemicals. It takes a realistic approach to this goal: The chemicals are to be phased out over time, allowing the development and implementation of less harmful replacements, and different phase-out schedules are set for developing and developed countries. These schedules are the result of extensive negotiation and debate and represent targets that can actually be reached.

Article 8 of the Montreal Protocol provides that "the Parties, at their first meeting, shall consider and approve procedures and institutional mechanisms for determining non-compliance with the provisions of this Protocol and for treatment of Parties found to be in non-compliance." The noncompliance procedures were made the responsibility of the protocol's implementation committee; they were modified in 1998 and provide that in the event of noncompliance three types of measures may be taken: (1) The non-complying party may be given "[a]ppropriate assistance, including assistance for the collection and reporting of data, technical assistance, technology transfer and financial assistance, information transfer and training"; (2) the implementation committee may issue a caution to the noncomplying party; (3) specific rights and privileges under the protocol may be suspended. Because the protocol provides for technological and financial assistance to developing countries, including transfer of technology, this last sanction actually has some teeth.

Especially when considered in contrast to the failure of international efforts to control greenhouse gas emissions, the Montreal Protocol can be considered a success. This does not mean that the problem is solved; a great quantity of ozone-depleting chemicals has already been released into the troposphere (the lower atmosphere). These chemicals will continue to find their way into the stratosphere for some time, and not much can be done about it. But the United Nations Industrial Development Organization states that according to "current estimates the CFC concentration in the ozone layer is expected to decline to pre-1980 levels by 2050" (Yalcindag, 2004).

Climate Change

International environmental law has had more success in addressing transboundary air pollution and ozone depletion than in addressing greenhouse gas emissions, although the Kyoto Protocol's entry into force is an encouraging sign.

Nearly all of the Earth's heat comes from the sun. The insolation (energy received from sunlight) of the Earth's upper atmosphere is about 1.4 watts per square meter. Of this amount, about half or a bit more reaches the surface of the earth; about 20 percent is absorbed by the atmosphere, and 25–30 percent is reflected back into space by clouds and other particles in the atmosphere. About 15 percent of the energy that reaches the surface is reflected back into space, much of it by highly reflective ice and snow surfaces such as that covering Antarctica. Of the energy absorbed by the Earth's surface, some is later re-emitted as lower-frequency infrared radiation. Certain gases in the atmosphere are relatively transparent to the incoming solar radiation but opaque to the re-emitted infrared radiation. These gases are known as greenhouse gases; the greater the amount of greenhouse gases in the atmosphere, the greater the amount of energy that will be retained. The Earth's current average surface temperature (about 59 degrees Fahrenheit) is the result of greenhouse gases; were there no greenhouse gases present, more heat would be radiated into space and the temperature would instead hover just above zero (Ma, 2004).

An increase in the amount of heat retained by the earth's atmosphere has the potential for disastrous environmental effects. The possibility of a world flooded by the melting of the polar ice caps has captured the popular imagination, but in the near term increases in sea level are more likely to result from thermal expansion of the water in the ocean. And there is a popular misconception that every place on earth will simply become a couple of degrees warmer, reflected in humorous bumper stickers along the lines of "Alaskans for Global Warming."

The Earth's atmosphere and hydrosphere are dynamic, not static; they can be viewed as an enormous heat engine, constantly moving quintillions of tons of air and water around the planet. Heat in this system is expressed as motion; more heat means more motion. The most immediate and drastic result of an increase in the heat retained by the earth will not be a few more beach days; it will be an increase in weather activity: more storms, more wind, more hurricanes, and tornadoes. A single hurricane can cause billions of dollars in economic damage; even a small increase in the total number of hurricanes could utterly destroy the economies of tropical countries and severely affect even developed countries such as the United States.

In the worst-case scenario, the emission of greenhouse gases becomes subject to self-reinforcing feedback mechanisms. Arctic ice, for example, contains clathrate methane, a greenhouse gas; as the ice melts, it releases the methane, which accelerates the greenhouse effect. And water vapor is a greenhouse gas; the heating of water increases the amount of water vapor in the atmosphere, which increases the amount of heat retained. When surface water temperatures rise above 80 degrees Fahrenheit, runaway feedback becomes possible; it is probable that a runaway greenhouse effect of this sort occurred on the planet Venus, leaving it with surface temperatures hot enough to melt lead.

Many scientists doubt that such a runaway greenhouse effect is possible on Earth; homeostatic mechanisms serve to counteract the feedback mechanisms. When ocean temperatures rise to the critical temperature, storm systems form, distributing the heat. And more water vapor leads to more clouds, which reflect sunlight before it reaches the surface. Only one spot on Earth—the "warm pool" in the Pacific Ocean northeast of Australia—consistently exhibits the conditions necessary for runaway feedback. And because the affected area is quite small, the excess heat can be distributed to other areas of the ocean and atmosphere. The upper atmosphere of Venus, after all, receives far more insolation than the upper atmosphere of the Earth. Venus receives about 2.6 watts per square meter to the Earth's 1.4 watts per square meter, or nearly twice as much. Pessimists point out, though, that the albedo (the ratio of reflected to incident light) of Venus is much higher than that of Earth: Most of the sun's light is reflected from the permanent clouds that shroud Venus, so that the planet only absorbs about 63 percent as much solar radiation as the Earth. The present-day temperatures on Venus are not due to the higher insolation, but due to the higher concentration of greenhouse gases in the atmosphere. The higher insolation may have caused the greenhouse effect—but astronomers speculate that at the time Venus suffered its runaway greenhouse effect, the insolation may have been only a third higher than, rather than twice as high as, Earth's is today. Both the cloud cover that reflects light back into space and keeps Venus cooler than it otherwise would be and the greenhouse gases that retain what heat reaches the surface may be the result of a runaway greenhouse effect (Fogg, 1995;

Weissman et al., 1999). Runaway feedback is thus something of a misnomer; the system eventually reached equilibrium, at a surface temperature of over 800 degrees Fahrenheit.

Even without such apocalyptic consequences, global warming is a problem that bears watching. Some scientists question whether global warming is actually occurring; what seems incontrovertible, however, is that the concentration of certain greenhouse gases in the atmosphere has increased greatly in the last two centuries as a result of human activity. The dual tasks of international law in this area are thus to monitor greenhouse gas levels and, more controversially, to attempt to reduce them.

Climate change is the subject of the United Nations Framework Convention on Climate Change (UNFCCC), which entered into force in 1994, and its offshoot, the Kyoto Protocol, which entered into force on 16 February 2005. That the Kyoto Protocol entered into force at all surprised many observers; by early 2004 it was widely believed to have been doomed by the failure of two major producers of greenhouse gases, the United States and Russia, to ratify.

The UNFCCC required its parties to reduce their emissions of greenhouse gases to 1990 levels by the year 2000 and imposed a general obligation on developed countries to make financial and technological transfers to developing countries to assist them in meeting these goals. All parties were required to develop inventories of greenhouse gases and to create national programs for mitigation and adaptation. These requirements were set in general terms; it was anticipated that a more specific plan for greenhouse gas reduction, with specific requirements for each of the member states, would be set in an additional protocol or protocols.

The incremental framework-plus-protocol approach does not always work, however, and in the case of global warming it very nearly failed completely. The Kyoto Protocol, by its terms, could not come into force until it was ratified or otherwise accepted by at least fifty-five countries, representing at least 55 percent of the world's emissions of greenhouse gases. In 2001 negotiations between the United States and those states that had already ratified or otherwise accepted the Kyoto Protocol broke down, and the United States withdrew from the Kyoto process. Although this breakdown was widely blamed on the intransigence of the Bush Administration, the U.S. Senate had expressed grave reservations about the Kyoto Protocol as early as 1997.

At this point the only non-party (other than the United States) with the ability to bring the Kyoto Protocol into force was Russia. Russia is a major producer of greenhouse gases, but Russian representatives had expressed fears that curbing production of those gases would curtail Russian industry and economic growth, keeping Russians, in the words of one Russian diplomat, "mired in poverty" (*Economist,* 2004). An alternate view was that because Russia's Kyoto Protocol targets for greenhouse gas reduction were based on (higher) Soviet-era levels of production, they might in fact have already been met. According to this line of reasoning, Russia had expected to receive a windfall from the Kyoto Protocol by selling the emission credits for reductions in excess of its target levels to other countries that had not met their targets. The withdrawal of the United States from the process greatly reduced the value of those credits, and Russia's subsequent resistance to the Kyoto Protocol now might have been a bargaining tactic aimed at increasing the price the European Union countries were willing to pay or aimed at obtaining other favors such as European support for Russian membership in the World Trade Organization (*A Green Future 2004; Climate Change: Carry On Kyoto 2004*). Whatever the reason, however, in November 2004 Russia ratified the protocol; by its own terms, the protocol took effect ninety days later.

In the meantime the European Union has created its own internal emissions trading market, opening in 2005. A market-based emissions trading program takes advantage of the fact that greenhouse gases are a global, not local, problem, and makes it possible to achieve the greatest total reduction in greenhouse gases at the lowest cost. Suppose, for example, that country A and country B were each required to reduce their greenhouse gas emissions. Country A might be able to reduce its emissions at a cost of $1 billion per X tons of greenhouse gas emission reduction. Country B might be able to receive the same goal at a cost of $4 billion. Because greenhouse gas distribution is global rather than local, the world as a whole is not better off if countries A and B each reduce their emissions by X than it is if country A reduces its emissions by 2X, and country B doesn't reduce its emissions at all; the net result is the same, but the total cost of the first option is $5 billion, and the total cost of the second option is only $2 billion. In fact, country A could receive a reduction of 5X for the same cost as countries A and B together achieving a reduction of 2X. It is thus more efficient, both economically and environmentally, for country B to pay country A to reduce emissions than for country B to reduce its own emissions.

The problem is that the world's largest producer of greenhouse gases, the United States, is still not a party to the Kyoto Protocol. It thus has no need to buy emissions credits, which decreases the value of those credits. In addition, there may be (with the possible exception of Russia) no sellers: Many industrialized countries have already failed to adhere to the emissions reductions to which they had agreed under the UNFCCC, and greenhouse gas emissions worldwide are continuing to increase. According to current projections, by 2010 the greenhouse gas emissions of the developed countries (excluding Russia) that have ratified or otherwise accepted the Kyoto Protocol will exceed the protocol's target levels by 29 percent (Guruswamy, 1997).

The Kyoto Protocol itself anticipated future rule-making dealing with problems of enforcement: "procedures and mechanisms" for addressing noncompliance with "binding consequences" were to be adopted by amendment of the protocol. But while the protocol's entry into force is an encouraging sign, it has given rise to a new dispute: As long as the United States is not a party, meaningful reduction of greenhouse gas emissions is likely to be difficult or impossible. The rest of the world is thus likely to blame the United States (and perhaps Australia, another non-party) for the adverse effects of increased greenhouse gas concentrations in the atmosphere. It may be that even with U.S. participation, however, the Kyoto Protocol would be too little, too late: The 5.2 percent reduction (by 2012, for industrialized countries) it calls for is far less than the 60 percent reduction that some experts believe is necessary to avoid catastrophic climate change (BBC News, 2004).

The Kyoto Protocol is likely to remain unpalatable to the United States, however, as long as it fails to take economic and demographic realities into account. National economies and populations are not static, nor do they all change at the same rate. The United States has an expanding economy and an increasing population. Japan and the countries of the European Union have lower rates of economic growth and population increase, and their populations are aging more rapidly than that of the United States. Many European countries are actually experiencing population decrease, and Japan is now in its second decade of recession. These economic and demographic factors should make it easier for these countries to comply with their

Kyoto Protocol targets. (Even so, as we have seen, they seem not to have been able to do so.) For the United States to accept targets based on similar assumptions would require the United States to bear the heaviest burden under the Protocol.

Biodiversity and Biopiracy

The term "biodiversity" encompasses three related but distinct concepts: the diversity of species, the diversity of ecosystems within a region, and the genetic diversity of individuals within a species.

No living creature, and no species of living creature, exists in isolation; all depend on other creatures for survival and are depended upon by others in turn. Tear too many holes in this web of life, and it falls apart; the result is a worldwide ecological collapse along the lines of the Permian extinction.

Even in the absence of global catastrophe, biodiversity has value. The genetic diversity of individuals within a species is essential to the survival of the species; non-diverse populations are vulnerable to disease or slight environmental changes that can be withstood by diverse populations. The diversity of species is important not only aesthetically and for the sake of the species themselves, but because so many species of plants and animals are essential to human survival—as food, for instance—or can produce tangible, often not yet realized economic benefits (as sources of medicines, for example). The benefits of a diversity of ecosystems are less tangible and are only beginning to be understood. Forests, for example, sequester carbon and replenish oxygen; wetlands filter water. And, of course, many ecosystems (Australia's Great Barrier Reef ecosystem, for example) have measurable economic value as tourist attractions.

Protection of biological resources was one of the first goals of international environmental treaty-making, although not surprisingly early treaties reflected conservationist rather than preservationist concerns. As early as 1911, the United States, Great Britain, Russia, and Japan created the Convention for the Preservation and Protection of Fur Seals (no longer in force, although other treaties now protect many seal species).

A few treaties deal with general worldwide preservation of biodiversity; perhaps the most important are the Convention on

Biological Diversity (CBD) and the Convention on International Trade in Endangered Species of Wild Fauna and Flora (CITES). A great many treaties deal with a particular aspect of the environment, such as wetlands, or with a particular type of organism, or with a particular region.

The threats to biodiversity are numerous; almost every area of human activity has the potential to reduce biodiversity. Areas of particular concern are agriculture, urban expansion, dams, tourism, introduction of exotic species, and overexploitation.

Agriculture requires the clearing of large quantities of land. A diverse population of organisms is replaced with a non-diverse monoculture; the total mass of living matter on the land may remain the same or even increase, but genetic diversity is lost. Agriculture also threatens biodiversity on land that it does not occupy through the harm caused by fertilizer and pesticide runoff.

Urban expansion also requires the clearing of large areas of land; the urban environment may be more biologically diverse than the average modern agribusiness farm, but it often destroys existing ecosystems to an even greater degree. Wetlands are drained and filled; forests are cleared; busy roadways present insurmountable barriers to animal migration, leaving those animals that remain as island populations too small to maintain their numbers and genetic diversity through reproduction, even if the resources upon which they depend for food have not been destroyed. And, of course, cities produce prodigious quantities of waste, which may poison ecosystems far beyond the limits of the city.

Dams are walls built across living rivers; they prevent the migration, essential to reproduction, of anadromous and catadromous fish. They flood large areas, creating lakes that become breeding grounds for mosquitoes and other disease vectors. They may alter water temperature and salinity, rendering the river uninhabitable for some of its previous inhabitants.

Tourism is a special threat to certain fragile yet frequently visited ecosystems, such as coral reefs. Oil discharged from tourist boats, or even sunscreen lotion from divers, can kill reef organisms. Boats accidentally collide with and damage reefs; dynamiting portions of the reef to make channels so that boats will not collide with the reef causes even more damage. Boats anchors also damage reefs. Coral and reef organisms are often over-harvested to be sold as souvenirs. Runoff from the construction of hotels may suffocate the reef; if it does not, untreated sewage from the operation of the hotels may. Eventually the reef ecosystem that attracted the tourists in the first place may be largely or entirely dead. The sad irony is that many of the tour operators exercising unsound practices in bringing tourists to the reef bill themselves as offering "ecotourism," a now terribly overused label.

Introduction of exotic species is a worldwide problem. A single insect hitching a ride in a box of fruit may ultimately lead to billions of dollars in crop damage in the country in which it arrives. A pet fish from an aquarium, released into the wild, may transmit a lethal fungus to an entire species of wild fish. The Nile perch, deliberately introduced into Africa's Lake Victoria, has completely destroyed the lake's preexisting ecosystem, virtually exterminating the indigenous cichlids. This, in turn, has had broader repercussions: The inhabitants of the lake basin, who formerly dried and stored the meat of the cichlids for food, must now smoke the fattier meat of the Nile perch in order to store it. To do this they have, over time, cut down much of the forest around the lake, leading to increased erosion and higher silt levels in the lake, as well as damaging the forest ecosystem (Goldschmidt, 1996).

Many species of animals, on land and in the sea, and some plants have been pushed to or near the brink of extinction by over-harvesting. Whales, elephants, and rhinoceroses survive today largely or entirely because of diligent conservation efforts. Many of the world's fisheries have been depleted to the point that it is no longer worthwhile to exploit them.

The threats to biodiversity cover too broad a range to be neatly addressed by any single treaty. The problem cannot be neatly compartmentalized, the way "Antarctica" or "ozone depletion" might be. Instead, protection of biodiversity has proceeded piecemeal, on a region-specific or issue-specific basis.

An early attempt to address global biodiversity loss, based on incomplete scientific understanding, was CITES. CITES aims to protect species that are especially endangered by international trade, such as elephants (hunted for ivory) or macaws (hunted for sale to pet shops). CITES restricts or prohibits trade in certain animal and plant species that are in danger of extinction. Nearly all of the world's countries are parties to CITES, and it can probably be credited with saving many species from extinction. But it has been subject to criticism on three grounds. First, many species extinctions are the result, not of trade, but of other factors, such as habitat loss or pollution. Second, CITES has been attacked by some environmentalists for its focus on "charismatic megafauna" —elephants, pandas, and the like—although this focus is more or less inevitable given that the purpose of CITES is to protect commercially viable species from harm caused by trade. Third, many or most of the species protected by CITES come from the developing countries of the South; nearly all of the potential ultimate buyers for these species or products derived from them are in the North. Thus CITES, with the best intentions in the world, acts as one more barrier against southern access to northern markets.

A later, more comprehensive, but also less successful attempt to address biodiversity loss is the Convention on Biological Diversity (CBD). In 1987, UNEP created an ad hoc working group to consider a possible "umbrella convention to rationalize current activities"—that is, to consolidate the existing treaties into a workable whole, eliminating overlap and identifying lacunae. The umbrella convention envisioned by UNEP proved unattainable; instead, the CBD, a framework treaty, was adopted. The COP of the CBD has the power to seek "appropriate forms of cooperation" with the executive bodies of other conventions related to biodiversity, but no more. Its provisions are almost entirely aspirational; the anticipation was that future meetings of the COP would add substantive provisions. For the most part this has not yet happened, although the first meeting of the COP did establish a "Clearing House Mechanism for Technical and Scientific Cooperation" to promote and facilitate scientific and technological cooperation; this has often been the first step in the creation of effective multilateral resource management regimes, although the taking of the first step is no guarantee of ultimate success.

The focus of attention on biodiversity has led environmentalists to stress the value and irreplaceability of genetic resources; one idea much touted by environmentalists and popularized in movies such as *Medicine Man* has been the potential of new medicines derived from these resources through bio-prospecting. At one time bio-prospecting was hailed by developed country environmentalists as an example of a sustainable use. Tropical and subtropical ecosystems are treasure troves of biodiversity; they can, in theory, provide new medicines, food crops, flowers, and ornamental plants, as well as genes for cross-breeding and improving already farmed species.

In reality bio-prospecting, and the derogatory term "bio-piracy" often attached to it, provides a cautionary tale. So far, at least, more human energy has been devoted to the dispute over rights to tropical and subtropical genetic resources than those resources have turned out to be worth. In the pharmaceutical field, in particular, investment in bio-prospecting has yielded little reward. At the same time it has generated a backlash in the developing world.

Bio-prospecting can be done the hard way, by going out into the wilderness and testing plants and animals more or less at random to see whether they might be commercially useful in some way. Or a shortcut can be taken: Ethno-botanists, cultural anthropologists, and other researchers can talk to the people who live within an ecosystem, learning about their medicines and food crops. The knowledge acquired by indigenous peoples over generations can, in theory, save researchers decades of searching.

In the developed world, once a new food crop has been bred, or a new drug has been developed, the breeder or developer will generally patent the invention. The patent-holder will then enjoy a monopoly for a set period of time (generally twenty years) on growing the crop or manufacturing the medicine and can collect royalties from others wishing to do so.

To many people in the developing world this seems like theft. The "inventor" has hitched a free ride on the knowledge of the indigenous people, who have received little or no compensation for sharing that knowledge. Yet the problem is difficult to remedy: If the indigenous peoples are to receive a share in the revenues, who, precisely, shall receive that share? The particular person who informed the researcher of the medicinal use of a particular plant? The descendants of the person who first discovered that use? The country or countries in which the plant is discovered or of which the indigenous people are citizens? Does it matter whether the knowledge is specialized, held only by a few people, or whether it is widely known and the person who informed the researcher of the use was the first person the researcher happened to ask? Should revenues from traditional knowledge be distributed to individuals at all or be put into trust for an entire tribe or ethnic group? If the latter, who is to administer the trust? These problems are probably not insurmountable, but they have only begun to be addressed.

The problem has been exacerbated by the inflated early expectations of high returns from bio-prospecting. Many activists who originally envisioned the creation of vast wealth built on a base of developing country indigenous knowledge have not yet adapted their expectations to reality: Little wealth has been created, and much money has been lost, in bio-prospecting (*Economist*, 1999). Existing international intellectual property law on the topic offers little help. International intellectual property agreements, such as the Agreement on Trade-Related Aspects of Intellectual Property Rights (TRIPS) deal with intellectual property from a "northern" perspective, focusing on traditional categories: patent, copyright, and trademark. Little or no provision is made for traditional knowledge. But companies in affected industries, such as the pharmaceutical industry, have been cautiously receptive to the idea of compensating indigenous people for valuable knowledge; they have not exhibited the sort of inflexible behavior seen in, for example, the music recording industry when confronted with the Internet.

International environmental law relating to biopiracy has undergone a change of direction. In 1983, the Food and Agricultural Organization (FAO), an agency of the United Nations, had declared that "plant genetic resources are a heritage of mankind [and] consequently should be available without restriction." This is unfavorable to the interests of the poor countries within the borders of which a majority of the world's land-based plant genetic resources are located. Resemblances can be seen to other areas: Britain's and France's insistence on "freedom of navigation" of a river running through someone else's country; the Outer Space Treaty's guarantee of equal access to geostationary orbits located directly above, not the rich countries of the North, but the poor countries located on the equator.

The South has an increasing voice in the international rule-making process, however. As a result, the term "common heritage" was dropped early in the process of negotiating the CBD and replaced with the term "common concern." The CBD rejects the idea that there should be free access for all the world to the genetic resources of a developing country. Article 15(1) of the CBD provides that "the authority to determine access to genetic resources rests with the national governments and is subject to national legislation." Article 15(5) adopts a rule of prior informed consent, permitting developing countries to negotiate a price for access to genetic material. (The CBD's sixth COP developed these ideas in considerably greater detail and expressed them in the non-binding Bonn Guidelines on Access to Genetic Resources and Fair and Equitable Sharing of the Benefits Arising out of their Utilization.) That the FAO has now adopted this approach as well can be seen in the FAO's International Treaty on Plant Genetic Resources for Food and Agriculture. At present, then, international law has managed to create a regime to provide payment for a resource that can be exploited sustainably but that seems at present to be of little actual value.

Other Areas of Dispute

Disputes arise in every area of international law; often, although not always, these disputes express aspects of the North-South conflict. In the brief examples below these disputes are addressed in several ways. In the case of toxic waste disposal, a group of developing countries in Africa banded together to resist a treaty regime that unfairly favored the economic interests of the North and formed a treaty regime of their own; together they achieved what none could have achieved independently—a fundamental alteration in the global regime governing transboundary shipments of toxic and hazardous wastes.

The Gabčíkovo-Nagymaros dispute was a freshwater resource dispute; below is a brief discussion of the attempt, not yet successful, to codify existing law on freshwater resources, along with a similarly brief discussion of the ocean environmental regime. Both of these topics are discussed at length in other volumes in ABC-CLIO's Contemporary World Issues Series: *Freshwater Issues* and *Ocean Politics and Policy*.

Finally, a look at space law provides insight into the scope of international environmental law. Outer space lies beyond the biosphere and is not generally thought of as having an "environment"; certainly it has no ecology. But even in space there are resources, and where there are resources there are bound to be resource disputes. The existing regime allocating those resources was designed by the North and favors northern interests, yet one tiny southern country has been able to challenge that regime, with some success.

Disposal of Toxic and Hazardous Substances on Land

Humans live on the land, not in the air or the oceans. The Earth's land surface environment is under more direct and constant attack from human activities than the atmosphere or the oceans, yet it receives less attention in international law. Most of the environmental harm to the terrestrial environment takes place within national borders; when it is international, it generally is so because the harmful activity—such as agriculture, industry, waste dumping, or urban development—takes place on, near, or across an international boundary, and so two states are involved. Transboundary harm to the terrestrial environment affects multiple states in only a few cases, such as the waging of war by one state upon another or the loss of biodiversity—both discussed elsewhere in this volume—or in the case of transboundary shipment and disposal of hazardous wastes.

Major multilateral agreements addressing the transboundary movement of hazardous chemicals include the Basel Convention on the Control of Transboundary Movement of Hazardous Wastes and their Disposal (Basel Convention) and the Bamako Convention on the Ban of Imports into Africa and the Control of Transboundary Movement and Management of Hazardous Wastes within Africa (Bamako Convention).

Storage and disposal of toxic wastes is always controversial; everyone realizes the necessity, yet no one would voluntarily choose to live next to a toxic dump site or waste incinerator. The presence of such facilities lowers real estate values and drives out all who can afford to leave. And communities or entire states often wage fierce campaigns to prevent the location of such facilities in their territories. As a result, the facilities are often located in communities that are both poor and politically disempowered. In the United States, these are often minority communities; civil rights and environmental justice activists now publicize data on this "toxic racism."

The same problem exists internationally; wealthy, industrial countries ship waste to poorer countries that desperately need the money they receive for allowing their country to be used in this way. Prior to 1989, the then-independent country of East Germany was a major recipient of this "toxic tourism," much of it from West Germany. The reunification of the two Germanies has, with fitting irony, brought the waste disposal problem back to a government that thought it had gotten rid of it.

Some developed countries, such as the United Kingdom, receive certain types of toxic waste; in these countries environmentalist opposition campaigns tend to be well organized, waged both in court and in the media. The result is to significantly raise overall disposal costs, and, in some highly publicized cases, even to send the wastes back to their country of origin. At one time the desperately poor countries of West Africa were deemed good candidates for waste-disposal sites; toxic waste was sometimes shipped to West Africa in leaking drums that were simply dumped on the ground, in unlined landfills, and left to gradually corrode and drain their contents into the water table. These countries lacked effective environmental movements capable of successfully opposing waste shipments or seeing that safe dumping standards were enforced. They thus became, in effect, importers of environmental problems. The rich countries of the North became the exporters of environmental problems, just as Australia exported problems to Romania with the Aurul mine, or Austria sought to export problems to Hungary with the Nagymaros project.

While the Basel Convention's existing regime governing transboundary movement of toxic wastes may have served to protect the environmental interests of developed countries, it seemed to be failing the developing states of West Africa. Several African countries formed the Bamako Convention (Bamako is the capitol of Mali) to ban all import of hazardous wastes into the territory of the member states from non-parties, and restricting other waste shipments more severely than the Basel Convention had. The Conference of Parties of the Basel Convention has subsequently attempted to amend the convention to reflect these concerns; the Ban Amendment, not yet in force, would prohibit all shipment of hazardous wastes, other than for recycling, from Organization for Economic Co-operation and Development (OECD) and European Union (EU) states (plus Liechtenstein) to non-OECD, non-EU states (again, plus Liechtenstein).

Ordinarily the Basel Convention prohibits "hazardous wastes or other wastes to be exported to a non-Party or to be imported from a non-Party." The United States is not a party to the Basel Convention but has continued to export wastes to the territory of states that are parties to the convention under a provision that permits such transfers when they do not derogate from environmentally sound management and when the Basel Convention's secretariat is informed of the transfer.

Fresh Water

The development of the regime of customary international law governing protection of freshwater resources is described in some detail in the section of Chapter 1 dealing with the development of modern concepts of state responsibility for transboundary environmental harm. Today it is widely accepted that states have a responsibility for environmental harm to a transboundary watercourse that affects the territory of lower riparians or co-riparians. But our understanding of "watercourses" changed enormously as we learned more about hydraulic systems. We now know that surface watercourses are a relatively minor freshwater resource: The oceans contain 97.3 percent of the Earth's water. More than three-quarters of the remainder (77.2 percent of the remaining 2.7 percent) is frozen in the polar ice caps and glaciers. Of the 22.8 percent not frozen in the ice caps and glaciers, nearly all (22.4 percent) is groundwater. Rivers, lakes, and other surface water together contain only 0.36 percent of the Earth's freshwater, or less than 0.01 of 1 percent of the total amount of water; clouds and water vapor make up the remainder (Barberis, 1991).

As a result, it has become apparent that aquifers (containing groundwaters) are the world's most valuable freshwater resource. Surface waters provide uses that groundwaters do not: transport, fisheries, and recreation. But the future of agriculture, in particular, depends on careful aquifer management.

Rivers, lakes, aquifers, and other freshwater resources are by nature local or regional resources rather than global ones. There is thus a very large number of international agreements dealing with particular freshwater resources, as well as a large body of local custom developed among the states within a particular drainage basin.

The closest thing to a universal statement of the law regarding the environmental protection of transboundary watercourses is the United Nations Convention on the Law of the Non-navigational Uses of International Watercourses (UNCLNUIW), adopted by the General Assembly in 1997 but not yet in force. After years of effort on the part of the International Law Commission (ILC), the convention, in somewhat altered form, was adopted by a vote of 103 in favor to 3 against, with 27 abstentions and 33 members absent. (A more accurate vote count might have been 106 in favor, 26 abstaining, and 31 absent: The representative of Belgium later said that Belgium should have been recorded as in favor rather than abstaining, and the representatives of Fiji and Nigeria, both recorded as absent, later said that they had intended to vote in favor.) While few states voted against the convention, the relatively large number of abstentions and the misgivings expressed even by some of the states voting in favor of the convention raise doubts as to whether it is a completely accurate statement of pre-existing law on the topic.

In particular, the UNCLNUIW is subject to criticism on the grounds that it does not adequately take into account the concerns of developing countries that are upper riparians along international watercourses. In other words, it values downstream territorial integrity, and thus environmental protection, to a degree that may prevent poorer upstream countries from developing the water resources within their borders.

The UNCLNUIW is a framework convention in the post-Stockholm mode. Thus, it requires states to do little other than gather and share information and notify each other. Its language is largely hortatory, but it does require notification, consultation, and negotiation when states plan any watercourse-related activity that might adversely affect other riparian states. It also requires parties to "take all appropriate measures to prevent the causing of significant harm to other watercourse States," and "where significant harm nevertheless is caused to another watercourse State, [to] take all appropriate measures . . . to eliminate or mitigate such harm and, where appropriate, to discuss the question of compensation."

The Ocean

The ocean environment suffers both from pollution and from overexploitation of fisheries resources. Ocean problems can be local or regional, but are often (as in the case of vessel-based oil pollution or depletion of pelagic fisheries) global, or at least of concern to the entire world.

Under traditional environmental law, protection of the ocean—at least, of the high seas—presented a special problem: The high seas are not the territory of any state. If harming the ocean harms no state's territory, there is no breach of state responsibility under the traditional *Trail Smelter/Corfu Channel* formulation.

The acceptance of Principle 21 of the Stockholm Declaration in the practice of states, and the incorporation of similar language in countless agreements and aspirational documents since, has changed matters. It is now a generally accepted principle of customary international law that states have a duty not to allow their territory to be used for activities that will cause harm to areas outside the territorial jurisdiction of any state. The treaty regime that governs the protection of the world's ocean environment is complex. Sources fall into two general categories: global and regional. The preeminent global agreement that addresses, among other ocean-related concerns, the marine environment is the United Nations Convention on the Law of the Sea (UNCLOS). The UNCLOS is still largely, perhaps primarily, concerned with navigation. While the decline in river shipping over the past century has led to a consequent shift in emphasis in law regarding watercourses from freedom of navigation to environmental protection, no similar shift has occurred in ocean law. Navigation of the seas remains important, both commercially and militarily; environmental concerns have been added to navigation concerns rather than replacing them

In addition to UNCLOS, a great many treaties govern specific aspects of the marine environment or specific seas. Many of these treaties were reached under the auspices of UNEP and follow the COP rule-making model described in Chapter 1. But other agreements either establish intergovernmental organizations (IGOs) to carry out rule-making, enforcement, and dispute resolution functions, or assign those functions to already-existing IGOs. The Convention on Future Multilateral Cooperation in the Northwest Atlantic Fisheries, for example, establishes the Northwest Atlantic Fisheries Organization to carry out its functions; the Convention on the Conservation of Antarctic Marine Living Resources establishes the Commission on the Conservation of Antarctic Marine Living Resources; and the Convention for the Prevention of Marine Pollution from Land-based Sources establishes the Paris Commission, which continues to perform these functions under the Convention for the Protection of the Marine Environment of the North-East Atlantic, which replaced it. The functions of the International Convention for the Prevention of

Pollution from Ships and of the International Convention on Oil Pollution Preparedness, Response and Co-operation, on the other hand, are carried out by the International Maritime Organization.

Space

In the five decades that humanity has been exploring space, the near-Earth outer space environment has grown cluttered (relatively speaking) with bits of orbiting debris. These are an environmental problem chiefly to the extent to which they may interfere with or damage commercial and government satellites or spacecraft, although there is always the possibility that they may fall to Earth and cause harm.

A much greater worry is radioactive material in space; nuclear power plants used to power satellites might cause radioactive contamination of the Earth, especially if they were to disintegrate on re-entry into the Earth's atmosphere and distribute their contents over a large area. And certain natural resources in space may prove valuable; problems of allocation and depletion of these resources then arise.

The uses of outer space are governed to some extent by a treaty regime, the foundational document of which is the Treaty on Principles Governing the Activities of States in the Exploration and Use of Outer Space, Including the Moon and Other Celestial Bodies (Outer Space Treaty).

Viewed from a terrestrial environmental protection perspective, harm to the outer space environment is significant only insofar as it reaches the earth or interferes with other human activities for its own sake. The prospect of purely aesthetic or ecological harm to outer space at this point seems remote. It is possible that, for instance, an advertising company might wish to paint a corporate billboard across the entire visible surface of the Moon, and no doubt many would view such an action as an environmental atrocity. But such an event seems unlikely, although as far as ecological harm goes, concern has been expressed that Earth-based microorganisms might hitch rides on research vehicles to such places as Mars and Europa, both of which are believed to be capable of sustaining life. These microorganisms might then contaminate and disrupt any extraterrestrial ecosystems already existing (Space Studies Board, 2000).

But space law may also be viewed from a natural resources perspective: There is a large but finite number of commercial and government satellites that can be placed into orbit around the Earth in cislunar space (that is, space no further away from the Earth than the orbit of the moon). Therefore, orbits are a resource that can be depleted. Certain types of orbits are more valuable than others. A particularly unusual orbit is the geostationary orbit: A satellite orbiting above the equator at an altitude of 22,300 miles will move around the center of the planet at the same speed as the surface below it; to an observer on the surface below, the satellite will always be directly overhead. At one time this made geostationary orbits valuable as a location for communications satellites; the number of such satellites that can be placed into geostationary orbit is limited to the number that can be placed in orbit 22,300 miles above the equator without bumping into each other—in other words, the number that can be strung out along a line about 165,000 miles long. And within this category some geostationary orbits—those above populated areas—may become more valuable than others.

The countries that underlie these geostationary orbits—that is, the nations situated on the equator—are developing countries; a few may have the resources to launch satellites, but the countries most likely to make use of the geostationary orbits for commercial purposes are the developed countries of North America, Europe, and East Asia. With regard to geostationary orbits, at least, these are the countries whose interests are expressed in the Outer Space Treaty's provisions that outer space shall be "free for exploration and use by all states without discrimination of any kind, on a basis of equality," and not subject to any nation's "claim of sovereignty, by means of use or occupation, or by any other means." (Oddly, Article I also contains a provision that there shall be "free access to all areas of celestial bodies," but not to space itself, although the "freedom of exploration and use" might seem to cover the same ground.)

This "freedom of exploration and use" is a latter-day analogue of the "freedom of navigation" that the colonialist powers of the early twentieth century sought for international rivers such as the Danube. Non-riparians such as England and France wished to make use of the Danube for navigation and had the ability to impose their wishes upon the states in whose territory the Danube was actually located; the non-riparians thus gained something of value without compensating the riparian states.

Similarly, the developed nations of the late twentieth century wished to exploit geostationary orbits, even though these orbits were located over the territory of other states; the Outer Space Treaty permits them to do so. In 1976, eight equatorial states (Brazil, Colombia, the People's Republic of the Congo, Ecuador, Indonesia, Kenya, Uganda, and Zaire [now Congo]) took exception to this idea and issued the Bogotá Declaration claiming sovereignty over the geostationary orbits above their territories. Nothing much has come of this, although between 1988 and 1990, Tonga, a tiny island nation located within the tropics but well south of the equator in the Pacific, submitted filings to the International Frequency Registration Board (a specialized UN agency that manages geostationary orbit points) for sixteen geostationary orbits over the Pacific. Tonga's action prompted an outcry from the members of the International Telecommunications Satellite Cooperation (INTELSAT), an international consortium that owns and manages communications satellites. INTELSAT is dominated by states capable of launching satellites; it has a procedure to provide access for smaller and poorer nations, but Tonga chose to bypass INTELSAT entirely. After negotiation, Tonga agreed to a compromise by which it received rights to six of the sixteen orbital positions it had originally sought (Cahill, 2001).

Meanwhile, changes in telecommunications technologies have rendered geostationary orbits less valuable, although Tonga has continued to press controversial claims to other valuable orbits. But the unique feature of geostationary orbits—being stationary relative to a point on the Earth's surface—may again become valuable. The geostationary orbits would be the natural location for solar-power satellites or space elevators, should it ever become technologically feasible to build such structures. In all likelihood the countries with the resources to build them, though, will not be the countries underlying the geostationary orbits; in that case, the question of rights to those orbits will eventually have to be addressed again.

Summary

This overview of problems, controversies, and solutions in international environmental law looks at a variety of environmental problems. In most cases the underlying controversy in some way reflects the tension between environment and development, pitting the interests of developing countries against those of developed countries. The classic UN model of dispute resolution through formal proceedings before the International Court of Justice has not proven entirely successful, even when both parties are sincerely committed to the process; negotiation and treaty-making have had somewhat more success. The incremental approach of most post-Stockholm environmental treaties has been successful in addressing problems of ozone depletion and chemical and particulate air pollution but not in addressing global climate change. A comprehensive treaty regime governing Antarctica has also succeeded both in resolving the territorial dispute that inspired it and in protecting the continent's environment. And developing countries have successfully defended their interests in outer space and prevented their countries from becoming toxic waste dumps for the developed world.

Other problems, though, have acquired an emotional aspect and seem incapable of resolution. The conflict over agricultural trade pits the interests of many developing nations directly against the interests of a relatively small but influential industry in the developed world; the governments of developed countries, being democratic, are unable to act against the interests of their own farmers. And the conflict over biopiracy simmers on in the popular imagination, long after any realistic hope of substantial returns has faded. Despite these failures, however, the last few decades have, on balance, been relatively successful ones, with many conflicts being resolved by working solutions, if not permanent ones.

Sources and Further Reading

Books and Articles

Ball, Philip. *H20: A Biography of Water.* London: Weidenfeld and Nicolson, 1999.

Barberis, Julio. *The Development of International Law of Transboundary Groundwater.* 31 Natural Resources Journal 167 (1991).

Beatley, Timothy, et al. *An Introduction to Coastal Zone Management.* Washington, DC: Island Press, 1994.

Bhagwati, Jagdish. *In Defense of Globalization*. Oxford: Oxford University Press, 2004.

Bullock, Alan et al. *Great Rivers of Europe*. London: Weidenfeld and Nicolson, 1966.

Cahill, Susan. *Give Me My Space: Implications for Permitting National Appropriation of the Geostationary Orbit*. 19 Wisconsin International Law Journal 231 (2001).

Caron, David. *The Frog That Wouldn't Leap: The International Law Commission and Its Work on International Watercourses*. 3 Colorado Journal of International Environmental Law and Policy 269 (1992).

Charnovitz, Steve. *Free Trade, Fair Trade, Green Trade: Defogging the Debate*. 27 Cornell International Law Journal 459 (1994).

Cicin-Sain, Biliana, and Robert W. Knecht. *Integrated Coastal and Ocean Management: Concept and Practices*. Washington, DC: Island Press, 1998.

Climate Change: Carry On Kyoto. The Economist, 13 and 57, 9 October 2004.

D'Amato, Anthony, and Kirsten Engel, eds. *International Environmental Law Anthology*. Cincinnati: Anderson Publishing, 1996.

Davidson, Eric A. *You Can't Eat GNP: Economics As if Ecology Mattered*. Cambridge, MA: Perseus Publishing, 2000.

De Villiers, Marq. *Water: The Fate of Our Most Precious Resource*. Boston: Houghton Mifflin, 2000.

Dellapenna, Joseph W. *The Two Rivers and the Lands Between: Mesopotamia and the International Law of Transboundary Waters*. 10 Brigham Young University Journal of Public Law 213, 1996.

Diaz-Bonilla, Eugenio. *Can Rich and Poor Countries Agree on a Level Trading Field? New Research Quantifies Harm of Agricultural Subsidies and Protectionism*. International Food Policy. Research Institute Press Release, 27 March 2003. Available at http://www.ifpri.org/pressrel/2003/20030327.htm (visited September 27, 2004).

Durning, Alan Thein. *Guardians of the Land: Indigenous Peoples and the Health of the Earth*. Washington, DC: Worldwatch Paper No. 112, 1992.

Eckstein, Gabriel. *Application of International Water Law to Transboundary Groundwater Resources, and the Slovak-Hungarian Dispute over Gabčíkovo-Nagymaros*. Suffolk Transnational 19 Law Review 67 (1995).

Ethnobotany: Shaman Loses Its Magic. The Economist, 18 February 1999.

European Bank for Reconstruction and Development. *Environmental Impact Assessment Legislation: Czech Republic, Estonia, Hungary, Latvia, Lithuania, Poland, Slovak Republic, Slovenia 207*. London: EBRD, 1994.

Feshbach, Murray, and Alfred Friendly, Jr. *Ecocide in the USSR: Health and Nature Under Siege*. New York: Basic Books, 1992.

Fitzmaurice, John. *Damming the Danube: Gabčíkovo and Post-Communist Politics in Europe.* Boulder, CO: Westview Press, 1998.

Flavin, Christopher, and Odil Tunali. *Climate of Hope: New Strategies for Stabilizing the Earth's Atmosphere.* Washington, DC: Worldwatch Paper No. 130, 1996.

Fogg, Martyn J. *Terraforming: Engineering Planetary Environments.* Warrendale, PA: SAE Press, 1995.

Fraser, Andrew S., et al. *Water Quality of World River Basins.* Nairobi: UNEP, 1995.

French, Hilary F. *Green Revolutions: Environmental Reconstruction in Eastern Europe and the Soviet Union.* Washington, DC: Worldwatch Paper No. 99, 1990.

Galambos, Judit. "Political Aspects of an Environmental Conflict: The Case of the Gabčíkovo-Nagymaros Dam System." Perspectives of Environmental Conflict and International Relations, 75–76 (Jyrki Kakonen, ed., 1992).

Goldschmidt, Tijs. *Darwin's Dreampond: Drama on Lake Victoria.* Sherry Marx-Macdonald, tr., Cambridge, MA: MIT Press, 1996.

Graham, Edward M. *Fighting the Wrong Enemy: Antiglobal Activists and Multinational Enterprises.* Washington, DC: Institute for International Economics, 2001.

Guruswamy, Lakshman. *International Environmental Law in a Nutshell.* St. Paul, MN: West, 1997.

Handl, Gunther. *The International Law Commission's Draft Articles on the Law of International Watercourses General Principles and Planned Measures : Progressive or Retrogressive Development of International Law?* 3 Colorado Journal of International Environmental Law and Policy 132 (1992).

Hanqin, Xue. *Relativity in International Water Law.* 3 Colorado Journal of International Environmental Law and Policy 45 (1992).

Haq, Bilal U., et al., eds. *Coastal Zone Management Imperative for Maritime Developing Nations.* Dordrecht, The Netherlands: Kluwer, 1997.

Heywood, Peter, and Karoly Ravasz. *Danube Diversion Stirs Controversy.* Engineering News Record, 23, 9 February 1989.

Is Kyoto Dead? The United Nations Treaty on Climate Change Is in Less Trouble than It Looks. The Economist, 4 December 2003.

Jacques, Peter, and Zachary A. Smith. *Ocean Politics and Policy: A Reference Handbook.* Santa Barbara: ABC-CLIO, 2003.

Lammers, Johan G. *Pollution of International Watercourses.* Dordrecht, The Netherlands: Martinus Nijhoff, 1984.

Lejon, Egil. *Gabčíkovo-Nagymaros: Stare a Nove Hriechy.* Bratislava: Edicia Delta, 1994.

Liska, Miroslav B. *The Gabčíkovo-Nagymaros Project—Its Real Significance and Impacts.* Europa Vincet 6, 7 November 1992.

Logan, Bruce E. *Environmental Transport Processes.* New York: John Wiley and Sons, 1999.

Ma, Qiancheng. *Greenhouse Gases: Refining the Role of Carbon Dioxide.* Goddard Institute for Space Studies Science Brief, March 1998. Available at http://www.giss.nasa.gov/research/intro/ma_01/ (visited September 27, 2004).

Mann, Simon. *Angry Hungary Demands Compensation.* Sydney (Australia) Morning Herald, 8, 10 February 2000.

McCaffrey, Stephen C. *The Law of International Watercourses: Non-navigational Uses.* Oxford: Oxford University Press, 2001.

Middleton, Karen, and Sharon Kemp. *How It Happened.* The West Australian, 4, 10 February 2000.

Nollkaemper, André. *The Legal Regime of Transboundary Water Pollution: Between Discretion and Constraint.* Dordrecht, The Netherlands: Martinus Nijhoff, 1993.

Nostalgie de la Boue: Why France Is Still so Keen to Support and Protect a Declining Business. The Economist, 51, 29 May 2004.

Organization for Security and Co-operation in Europe. *OSCE Handbook, 3rd ed.* Vienna: OSCE Review, 2000.

Perczel, Karoly, and George Libik. *Environmental Effects of the Dam System on the Danube at Bos-Nagymaros.* 18 Ambio 247 (1989).

Petts, Geoffrey, and Peter Calow, eds. *River Biota: Diversity and Dynamics.* Oxford, UK: Blackwell Science, 1996.

———. *River Restoration.* Oxford, UK: Blackwell Science 1996.

Plater, Zygmunt B., et al. *Environmental Law and Policy: Nature, Law, and Society.* St. Paul, MN: West, 1992.

Postel, Sandra. *Last Oasis: Facing Water Scarcity.* New York: W.W. Norton, 1992.

———. *Dividing the Waters: Food Security, Ecosystem Health, and the New Politics of Scarcity.* Washington, DC: Worldwatch Paper No. 132, 1996.

———. *Pillar of Sand: Can the Irrigation Miracle Last?* New York: W.W. Norton, 1999.

Prokes, Jozef. *The Dam In Its True Light.* Europa Vincet 6, 12 November 1992.

Reibel, David Enrico. *Environmental Regulation of Space Activity: The Case of Orbital Debris.* 10 Stanford Environmental Law Journal 97 (1991).

Reynolds, Glenn H., and Robert P. Merges. *Outer Space: Problems of Law and Policy, 2nd ed.* Boulder, CO: Westview Press, 1997.

Rodríguez-Iturbé, Ignacio, and Andrea Rinaldo. *Fractal River Basins: Chance and Self-Organization.* Cambridge, UK: Cambridge University Press, 1997.

Roodman, David Malin. *The Natural Wealth of Nations: Harnessing the Market for the Environment.* New York: W.W. Norton, 1998.

———. *Paying the Piper: Subsidies, Politics and the Environment.* Washington, DC: Worldwatch Paper No. 133, 1996.

Schapiro, Mark. *The New Danube.* Mother Jones, 72, Apr./May 1990.

Schwabach, Aaron. *Diverting the Danube: the Gabčíkovo-Nagymaros Dispute and International Freshwater Law.* 14 Berkeley Journal of International Law 290 (1996).

———. *From Schweizerhalle to Baia Mare: The Continuing Failure of International Law to Protect Europe's Rivers.* 19 Virginia Environmental Law Journal 431 (2000).

———. *The United Nations Convention on the Law of Non-Navigational Uses of International Watercourses, Customary International Law, and the Interests of Developing Upper Riparians*, 33 Texas International Law Journal 257 (1998).

Serenyi, Juliet. *Danube Project Sours.* Christian Science Monitor, 19, 9 December 1992.

Smith, Zachary A., and Grenetta Thomassey. *Freshwater Issues: A Reference Handbook.* Santa Barbara: ABC-CLIO, 2003.

Space Studies Board. *Preventing the Forward Contamination of Europa.* 29 June 2000. Available at http://www7.nationalacademies.org/ssb/europapreface.html (visited September 26, 2004).

Sprankling, John G., and Gregory S. Weber. *The Law of Hazardous Wastes and Toxic Substances in a Nutshell.* St. Paul, MN: West, 1997.

Start Date Set for Kyoto Treaty. BBC News, 18 November 2004. Available at http://news.bbc.co.uk/go/pr/fr/-/2/hi/europe/4022283.stm (visited 18 November 2004).

Steinbeck, John. *The Grapes of Wrath.* New York: Viking Press, 1939.

Stiglitz, Joseph E. *Globalization and Its Discontents.* New York: W.W. Norton, 2002.

Tang, Xiyang. *Living Treasures: An Odyssey through China's Extraordinary Nature Reserves.* New York: Bantam, 1987.

Teclaff, Ludwik A. *Fiat or Custom: The Checkered Development of International Water Law.* 31 Natural Resources Journal 45 (1991).

Triggs, Gillian D., ed. *The Antarctic Treaty Regime: Law, Environment and Resources.* Cambridge, UK: Cambridge University Press, 1987.

Turner, Barry, ed. *The Statesman's Yearbook: The Politics, Cultures, and Economies of the World, 139th ed.* Houndmills, Basingstoke, UK: Palgrave, 2001.

United Nations Environment Programme, Office for the Co-ordination of Humanitarian Affairs. *OCHA, Cyanide Spill at Baia Mare, Romania: Spill of Liquid and Suspended Waste at the Aurul S.A. Retreatment Plant in Baia Mare.* Available at www.mineralresourcesforum.org/incidents/ BaiaMare/docs/final_report.pdf (visited 26 September 2004).

Vandevelde, Kenneth. *International Regulation of Fluorocarbons.* 2 Harvard Environmental Review 474 (1978).

———. *Investment Liberalization and Economic Development: The Role of Bilateral Investment Treaties.* 36 Columbia Journal of Transnational Law 501 (1998).

Vargha, Janos. *Egyre tavolabb a jotol.* Valosag, November 1981 (in Hungarian).

Viessman, Warren Jr., and Gary L. Lewis. *Introduction to Hydrology, 4th ed.* New York: Harper Collins, 1996.

Weissman Paul R., et al., eds. *Encyclopedia of the Solar System.* San Diego: Academic Press, 1999.

Wescoat, James L. Jr. *Beyond the River Basin: The Changing Geography of International Water Problems and International Watercourse Law.* 3 Colorado Journal of International Environmental Law and Policy 301 (1992).

Williams, Paul R. *International Environmental Dispute Resolution: The Dispute Between Slovakia and Hungary Concerning Construction of the Gabčíkovo and Nagymaros Dams.* 19 Columbia Journal of Environmental Law 1 (1994).

World Commission on Dams. *Dams and Development: A New Framework for Decision-Making.* London: Earthscan Publications, 2000.

Yalcindag, H. Seniz. *United Nations Industrial Development Organization, Service Module 7: Montreal Protocol-Overview,* undated. Available at http://www.unido.org/doc/5072 (visited September 26, 2004).

Zacklin, Ralph, and Lucius Caflisch, eds. The Legal Regime of International Rivers and Lakes. In *Le regime juridique des fleuves et des lacs internationaux 203.* Dordrecht, Netherlands: Martinus Nijhoff, 1981.

Treaties and Other International Agreements

African Convention for the Conservation of Nature and Natural Resources, 15 September 1968. 1001 U.N.T.S. 3.

Agreement Concerning the Establishment of a River Administration in the Rajka-Gonyu Sector of the Danube, 27 February 1968. Czechoslovakia-Hungary, 50 U.N.T.S. 640.

Agreement Governing the Activities of States on the Moon and Other Celestial Bodies. G.A. Resolution 34/68, UN GAOR, 34th Session, Supplement No. 46. U.N. Doc. /68 A/RES/3477(1979), 18 I.L.M. 1434 (1979).

Agreement on Air Quality, U.S.-Canada, 13 March 1991. 30 I.L.M. 676.

Agreement on Trade-Related Aspects of Intellectual Property Rights, 15 April 1994. 33 I.L.M. 1197.

Agreement to Cooperate in the Solution of Environmental Problems in the Border Area, U.S.-Mexico, 14 August 1983. 22 I.L.M. 1025.

Antarctic Treaty, 1 December 1959. 12 U.S.T. 794, 71 U.N.T.S. 402.

Bamako Convention on the Ban of Imports into Africa and the Control of Transboundary Movement and Management of Hazardous Wastes within Africa, 30 January 1991. 30 I.L.M. 773.

Basel Convention on the Control of Transboundary Movement of Hazardous Wastes and their Disposal, 22 March 1989. 28 I.L.M. 657.

Cartagena Protocol on Biosafety to the Convention on Biological Diversity, 29 January 2000. 39 I.L.M. 1027.

Convention for the Conservation of Antarctic Seals, 11 February 1972. 29 U.S.T. 441, 11 I.L.M. 251.

Convention for the Preservation and Protection of Fur Seals, June 4, 1974. 37 Stat. 1542, 7 July 1911 (no longer in force). T.S. No. 564.

Convention for the Prevention of Marine Pollution from Land-based Sources. 13 I.L.M., 352.

Convention for the Prohibition of Fishing with Long Driftnets in the South Pacific, 24 November 1989, 29 I.L.M. 1454.

Convention on Biological Diversity, 5 June 1992. 31 I.L.M. 818.

Convention on the Conservation of Antarctic Marine Living Resources, 20 May 1980. T.I.A.S. No. 10240, 19 I.L.M. 841.

Convention on International Trade in Endangered Species of Wild Fauna and Flora, 3 March 1973. 993 U.N.T.S. 243, 27 U.S.T. 1087, T.I.A.S. No. 8249, 12 I.L.M. 1085.

Convention on Long-Range Transboundary Air Pollution, 13 November 1979. 1302 T.I.A.S. No. 10541, U.N.T.S. 217, 18 I.L.M., 1442.

Convention on the Conservation of European Wildlife and Natural Habitats, 19 September 1979. E.T.S. 104.

Convention on the Continental Shelf, 10 June 1964. 15 U.S.T. 471, 499 U.N.T.S. 311.

Convention on the Prevention of Marine Pollution by Dumping of Wastes and Other Matter, 29 December 1979. 1046 U.N.T.S. 120.

Convention on the Protection of the Marine Environment of the Baltic Sea Area, 22 March 1974. 13 I.L.M. 546.

Convention on the Protection of the Rhine, 12 April 1999. Available from www.iksr.org/GB/bilder/pdf/convention_on_tthe_protection_of_the_rhine.pdf (visited 27 September 2004).

Convention on Wetlands of International Importance Especially as Waterfowl Habitat, 2 February 1971. 996 U.N.T.S. 245. Reprinted in 11 I.L.M. 963 (1972).

Convention Relating to the Protection of the Rhine Against Pollution by Chlorides, 3 December 1976. 16 I.L.M. 226.

Declaration of the First Meeting of the Equatorial Countries, 3 December, 1976. I.T.U. Doc. WARC-BS 81-E, (Bogotá Declaration). Reprinted in Nicolas Matte, Aerospace Law: Telecommunications Satellites (1982): 341-44.

FAO International Treaty on Plant Genetic Resources for Food and Agriculture. Available at http://www.fao.org/Legal/TREATIES/033s-e.htm (visited May 28, 2004).

Final Act Relating to the Establishment and Operation of the Iron Gates Water Power and Navigation System on the River Danube, Romania-Yugoslavia, 30 November 1963. 512 U.N.T.S. 6 [hereinafter Final Act] (English text begins at 512 U.N.T.S. 12; the Final Act is one of the twelve treaties regarding the Iron Gates project contained in volumes 512 and 513 of the U.N.T.S.).

International Convention for the Prevention of Pollution from Ships (MARPOL), 2 November 1973. 12 I.L.M. 1319.

International Convention for the Regulation of Whaling. 161 U.N.T.S. 72.

International Convention on Civil Liability for Oil Pollution Damage, 29 November 1969. 9 I.L.M. 45, 973 U.N.T.S. 3.

International Convention on Oil Pollution Preparedness, Response and Cooperation. I.L.M. 773, 30 November 1990 (1991).

International Convention on the Establishment of an International Fund for Compensation for Oil Pollution Damage, 18 December 1971. 1971 UN Jur. Y.B., 103.

International Convention Relating to Intervention on the High Seas in Cases of Oil Pollution Casualties, 26 November 1969. 970 U.N.T.S. 211, 26 U.S.T. 765.

Kuwait Regional Convention for Co-operation on the Protection of the Marine Environment from Pollution, 24 April 1978. 1140 U.N.T.S. 133, 17 I.L.M. 511.

Kyoto Protocol to the United Nations Framework Convention on Climate Change, 10 December 1997 (in force Feb. 16, 2005). 37 I.L.M. 22.

London Agreement on the Gabčíkovo-Nagymaros Project, 28 October 1992. Czechoslovakia-European Commission-Hungary, 32 I.L.M. 1291.

Montreal Protocol on Substances that Deplete the Ozone Layer, 16 September 1987. 26 I.L.M. 1550 (1987).

Protocol on Environmental Protection to the Antarctic Treaty, 30 October 1991. 4 I.L.M. 1461 (1991).

Rotterdam Convention for the Prior Informed Consent Procedure for Certain Hazardous Chemicals and Pesticides in International Trade, 10 September 1998.

Special Agreement for Submission to the International Court of Justice of the Differences Between Them Concerning the Gabčíkovo-Nagymaros Project. 7 April 1993, Hungary-Slovakia, 32 I.L.M. 1293.

Treaty Concerning the Construction and Operation of the Gabčíkovo-Nagymaros System of Locks, 16 September 1977. Czechoslovakia-Hungary, 1109 U.N.T.S. 235, 32 I.L.M. 1247.

Treaty on Principles Governing the Activities of States in the Exploration and Use of Outer Space, Including the Moon and Other Celestial Bodies, 10 October 1967. 610 U.N.T.S. 205, 6 I.L.M. 386.

United Nations Convention on the Law of the Non-navigational Uses of International Watercourses. G.A. Res. 51/229, U.N. GAOR, 51st Sess., 21 May 1997; 36 I.L.M. 700 (1997).

United Nations Convention on the Law of the Sea, 10 December 1982. UN Document A/CONF.62/122, 21 I.L.M. 1261.

United Nations Economic Commission for Europe Convention on the Protection and Use of Transboundary Watercourses and International Lakes, 17 March 1992. 31 I.L.M. 1312.

United Nations Framework Convention on Climate Change, 9 May 1992. 31 I.L.M. 849 (1992).

Vienna Convention for the Protection of the Ozone Layer, 22 March 1985. UNEP Doc. IG.53/5, 26 I.L.M. 1529 (1987).

Other International Materials

African Charter on Human and Peoples' Rights. OAU Doc. CAB/LEG/67/3/CAB/LEG/67/3 rev. 5, 21 I.L.M. 59 (1982).

Bonn Guidelines on Access to Genetic Resources and Fair and Equitable Sharing of the Benefits Arising out of their Utilization. U.N. Decision VI/24 CBD/Annex (2000). UN Document UNEP/CBD/COP/6/24 (2002).

Declaration of the Government of the Republic of Hungary on the Termination of the Treaty Concluded Between the People's Republic of Hungary and the Socialist Republic of Czechoslovakia on the Construction and Joint Operation of the Gabčíkovo-Nagymaros Barrage System, art. I(1) 32 I.L.M. 1260 (1993).

Gabčíkovo-Nagymaros Dispute (Slovakia v. Hungary), 1997 I.C.J. 7 (1997).

Johannesburg Plan of Implementation, September 2002. UN Document. A/CONF.199/20, 4.

Non-Binding Authoritative Statement of Principles for a Global Consensus on the Management, Conservation and Sustainable Development of all Types of Forests, 14 August 1992. UN Document A/CONF.151/26 (Vol. III).

Trail Smelter Case (U.S. v. Can.), 3 R.I.A.A., 1905, 1965 (1941). Reprinted in 35 American Journal of International Law 684 (1941).

3

Special Issues
for the United States

The United States is a developed country with a great wealth of natural resources, a high volume of environmentally significant trade with other countries, an active environmental movement, and perhaps the world's most thoroughly developed body of environmental law. It is also the world's leading military power and is frequently involved in armed conflicts.

This chapter looks first at the application of international law in the courts of the United States, as well as the problems of applying foreign law in U.S. courts and access to the U.S. court system by foreign plaintiffs. It then examines the inverse problem: the application of U.S. laws to activities outside the United States. Several major environmental statutes and court decisions discussing their extraterritorial application are discussed.

The chapter examines three international environmental disputes in detail: the litigation on behalf of indigenous peoples on the island of New Guinea against a U.S. mining company; U.S. barriers to agricultural trade, which have harmful effects on the unique Florida Everglades ecosystem; and two issues arising from the involvement of the United States in a war outside its territory—the 1999 Kosovo conflict. The first of these, the dispute over the bombing of a chemical manufacturing and storage complex, involves three separate dispute-resolution tools: litigation, fact-finding, and criminal investigation. The second is, like opposition to bio-prospecting and globalization, the result of incorrect information—the dispute over depleted uranium munitions, in large part arising from the incorrect belief that depleted uranium is highly radioactive.

International Law in U.S. Practice

Treaties

Treaties are ratified by the president upon receipt of the advice and consent of the Senate, the latter representing the sovereign states. Article VI, clause 2 of the U. S. Constitution provides that "All Treaties made, or which shall be made under the Authority of the United States, shall be the supreme Law of the Land; and the judges in every State shall be bound thereby, any Thing in the Constitution or Laws of any State to the Contrary notwithstanding." In practice this means that treaties supersede all inconsistent state law, both prior and subsequent. Treaty law also supersedes all prior federal law, but not subsequent federal law: the "last-in-time" rule provides that where a conflict exists between a treaty and a federal statute, whichever is most recent will prevail (Buergenthal, 2002; Trimble, 2002).

Treaties are either self-executing or non-self-executing. A self-executing treaty creates rights and obligations enforceable in the courts of the United States without any implementing legislation being passed by Congress; the text of the treaty itself is the law the court will apply. A non-self-executing treaty does not in and of itself create rights and obligations enforceable in U.S. courts; Congress must execute the treaty by enacting implementing legislation. In other words, a self-executing treaty is a mechanism by which the Senate and the president can create enforceable rights and obligations, but the Senate and House may then combine (by subsequent legislation) to alter or undo the effect of the treaty. A non-self-executing treaty is a mechanism by which the president and the Senate may suggest rights and obligations that ought to be created, but those rights and obligations are not enforceable until the House and the Senate combine to make them so (Bederman, 2001).

The actual boundary between self-executing and non-self-executing treaties is indistinct and the cause of much judicial uncertainty. Some theorists, especially in the human rights field, even criticize the idea of non-self-executing treaties, claiming that there is no constitutional basis for such a distinction; this view must be regarded as extreme, however.

The application of treaty law in the U.S. legal system thus requires first a determination of whether the treaty is self-executing; if it is, the court may treat it in the same way it would a statute. If the treaty is not self-executing, the next step is to determine whether implementing legislation has been passed; if so, that legislation can be applied, and if not the treaty has no force in domestic U.S. courts, even though it still creates binding rights and obligations between the United States and other states under public international law. In other words, a non-self-executing treaty might provide the basis for a suit by another state against the United States before the International Court of Justice (ICJ) but would not provide the basis for a similar suit, even by the same aggrieved foreign state, before a court in the United States under U.S. law. A country's internal law, in other words, is no excuse for non-performance of its duties to other countries under a treaty.

Custom

Customary international law presents a more difficult problem. U.S. courts are understandably reluctant to apply rules the very existence, let alone scope, of which is often in dispute. In a 1900 case, *The Paquete Habana*, the U.S. Supreme Court held that customary international law is "part of our law, and must be ascertained and administered by the courts of justice of appropriate jurisdiction, as often as questions of right depending upon it are duly presented for their determination." The *Paquete Habana* court, though, was confronting a situation in which there was no conflicting U.S. statute: "where there is no treaty, and no controlling executive or legislative act or judicial decision." Considerable controversy surrounds the situation in which customary international law is inconsistent with a prior federal statute. (A subsequent U.S. statute would be controlling under the "last-in-time" rule applied to treaties, and thus no controversy exists.) One often insoluble problem is determining when a particular practice became customary international law; the date of a treaty is easy to determine, but the date of a customary rule is not.

In addition to the question of applicable law, there is also the question of jurisdiction. Courts in the United States may be un-

able to deal with some disputes of an international character because they lack jurisdiction over the subject matter or over one or
more of the parties. The courts may lack subject matter because
of the political question and act of state doctrines and may lack
jurisdiction over a foreign governmental party under the Foreign
Sovereign Immunities Act (FSIA).

Under the political question doctrine, issues that are primarily political in character are inappropriate for judicial review; issues within the scope of the foreign affairs power are often political and, as the Supreme Court has stated, "uniquely demand
single-voiced statement of the government's views."

The act of state doctrine prohibits the decision by U.S. courts
of cases that call into question the sovereign acts of a foreign government. If the act were to be applied in an environmental context, the scenario might be something like this: A foreign government might decide to build a dam to generate electricity, and the
dam might flood property belonging to a U.S. party, rendering it
valueless. The U.S. party might then wish to proceed against the
foreign government in a U.S. court, seeking to attach assets of the
foreign government held in U.S. banks. Among the many obstacles the U.S. party would encounter would be the act of state
doctrine: The flooding of land to generate electricity, in effect
"taking" the flooded land, is an act within the sovereign power
of the foreign state. In a somewhat similar, although not environmental, case involving the nationalization of a U.S. sugar company by the government of Cuba, the U.S. Supreme Court stated
that "the Judicial Branch will not examine the validity of a taking
of property within its own territory by a foreign sovereign government, extant and recognized by this country at the time of
suit, in the absence of a treaty or other unambiguous agreement
regarding controlling legal principles, even if the complaint alleges that the taking violates customary international law." The
court also pointed out the close relationship between the act of
state doctrine and the political question doctrine: Involvement of
U.S. courts in such questions may interfere with the foreign policy efforts of the executive branch. And the act of state doctrine is
also closely related to (and often confused with) foreign sovereign immunity, as well.

The FSIA, enacted in 1976 to codify a pre-existing common-law doctrine, provides jurisdictional immunity to foreign
governments and their instrumentalities for their actions, with
three exceptions: Situations in which the foreign state has

waived its immunity, situations in which the foreign state is acting in a private or a commercial capacity, and situations involving property taken in violation of international law and now present in the United States in connection with a commercial activity of the foreign state. Note the difference between the act of state doctrine and immunity under the FSIA: the act of state doctrine deprives the courts of jurisdiction over certain types of subject matter (sovereign acts of a foreign state), as does the political question doctrine (foreign affairs and other political matters). The FSIA deprives courts of jurisdiction not over certain types of subject matter but over certain parties (foreign states).

The FSIA, the act of state doctrine, and the political question doctrine may operate simultaneously; a court may lack jurisdiction under two, or even all three, of these doctrines. In the environmental scenario described above, for example, the U.S. courts might lack jurisdiction over the subject matter under the act of state doctrine. If, for example, negotiations between the U.S. and foreign government to resolve the issue were taking place, the court might also lack jurisdiction over the subject matter under the political question doctrine. And, unless the foreign state waived its immunity, or the building of the dam and hydroelectric plant could be characterized as a commercial activity, the court would also lack jurisdiction over the foreign state as a party under the FSIA.

Foreign Law in U.S. Courts

The law of foreign countries is occasionally applied by courts in the United States. Choice of law is an area of legal study in its own right; each of the fifty states of the United States has its own rules regarding when and how it will apply foreign law in its own courts. As a general rule, courts in the United States will not apply the criminal or tax laws of foreign countries, although they may recognize judgments of foreign courts in these matters. In addition, courts in the United States will not apply provisions of foreign law that are contrary to U.S. law or public policy (for example, a foreign statutory provision restricting free speech or permitting racial discrimination), even when the state's rules regarding choice of law would otherwise dictate the application of the foreign law.

Extraterritorial Application of U.S. Environmental Laws

In the case of the cyanide spill at the Aurul gold mine at Baia Mare, Romania, discussed in the preceding chapter, Australian environmentalists were confronted with their country's complicity in an environmental disaster abroad and believed the disaster might have been prevented by the application of Australian law to the Australian party's actions. Environmentalists in the United States often have similar concerns; U.S. companies engage in practices overseas that might be restricted or prohibited in the United States on environmental grounds. The stock response to this concern is that if U.S. business is required to comply with U.S., rather than local, laws wherever it goes, it will be unable to compete with the businesses of other countries that do not impose such requirements on their nationals.

There is some truth to this. A similar instance in which the United States has long held the moral high ground is the problem of what is sometimes termed "payments to foreign government officials"—bribery. Since 1977, U.S. businesses have been forbidden, under the Foreign Corrupt Practices Act, to make corrupt payments to foreign officials for the purpose of obtaining or keeping business. This has put them at a disadvantage in competition with businesses from other developed nations; until recently, no other major trading nation had any law similar to the Foreign Corrupt Practices Act, and many even permitted corporations to deduct such payments from their corporate income taxes. Finally, after two decades of intense pressure from the United States, the other Organization for Economic Cooperation and Development (OECD) countries have agreed to the OECD Convention on Combating Bribery of Foreign Public Officials in International Business Transactions. The OECD Convention has been criticized for its many weaknesses—it does not even forbid its parties to provide tax deductions for bribes, for example (although most European countries no longer do so).

The amount of business lost by U.S. firms to foreign competitors as a result of bribery by those competitors is impossible to calculate with any degree of accuracy, but estimates tend to range from about $25 billion to about $50 billion per year, even after the OECD Convention (e.g., Taylor, 2001), although a few idealists insist, unrealistically, that no business at all has been lost. Thus U.S. companies may have sacrificed more than a trillion dollars' worth of business to the government's admirable attempt to stamp out global corruption. From a pragmatic viewpoint, the Foreign Corrupt Practices Act may be viewed as a failure. But from a moral standpoint, it has been a success: Decades of pressure from the United States, employing nothing more forceful than a moral argument, have brought about a change in the laws of most of the countries of the developed world and universal acknowledgment that it is wrong for countries to allow their citizens to bribe foreign government officials; a change in fundamental values has been achieved, and domestic legislation will follow. Ultimately, this moral success will be followed by economic success, since reducing corruption will save the world economy a great deal of money.

This same conflict between pragmatism and idealism exists in environmental law: Applying U.S. environmental laws to the foreign activities of U.S. businesses will almost certainly result in lost profits and lost business for as long as other developed countries do not follow suit. But few smaller economies can afford to take a similar position of moral leadership; unless the European Union as a whole agrees to do so, any initiative to apply developed-world environmental standards to investments and activities in developing countries can only come from the United States.

The Foreign Corrupt Practices Act is specifically aimed at the conduct of U.S. nationals and business entities outside the United States. Ordinarily there is a presumption that statutes enacted by Congress are not applicable to activities outside the United States, in the absence of some specific evidence in the statute itself to the contrary. At one time, this was an irrebuttable presumption. But the Supreme Court has gone through several stages in interpreting the extraterritorial application of statutes, beginning in the early 1900s. Currently, the law regarding the extraterritorial application of U.S. environmental laws is confused; there seems to be a presumption against it, but the extent of that presumption is unclear.

In the 1909 case of *American Banana Co. v. United Fruit Co.,* the U.S. Supreme Court refused to apply the Sherman Antitrust Act to actions committed in Costa Rica by one U.S. company to the detriment of another U.S. company, stating that in the absence of a clear indication to the contrary from the legislature, U.S. law should be "confined in its operation and effect to the territorial limits over which the lawmaker has general and legitimate power." Forty years later, in *Foley Bros., Inc. v. Filardo,* the Supreme Court addressed the question of what might constitute evidence of Congressional intent to have a statute apply extraterritorially. The plaintiff in *Foley,* Frank Filardo, was a U.S. citizen working overseas for a U.S. company performing work under contract to the U.S. government; Filardo claimed that his employer had violated the Eight-Hour Law, a federal labor law. The Supreme Court explained that a three-prong test must be applied to determine whether Congress intended the statute to be applied extraterritorially: First, there should be an express statement of such intent in the language of the statute. Second, if no such statement is found in the statute, it may be found explicitly or implicitly in the legislative history of the law (the various draft versions of the bill, committee reports, statements of members of Congress discussing the bill, and so forth). Third, such evidence may be found in administrative interpretations of the law. The Eight-Hour Law failed all three prongs of the test.

Between 1949 (after *Foley Brothers*) and 1990 the Supreme Court continued to wrestle with the problem. In 1991 the Supreme Court decided *Equal Opportunity Employment Commission v. Arabian American Oil Co.,* another employment case; this one alleged harassment (outside the United States) on the basis of race and religion by a U.S. employer against a U.S. employee. The majority applied the more stringent "clear statement of intent" test. Three justices dissented, however, and argued in a separate opinion that this high standard should apply only to situations with foreign policy implications; a labor dispute between a U.S. employee and his U.S. employer did not have such implications and did not infringe upon the sovereignty of other nations and the less stringent three-prong *Foley* test.

In an environmental context, the U.S. Court of Appeals for the District of Columbia Circuit addressed extraterritorial application of the National Environmental Policy Act (NEPA) in *Environmental Defense Fund, Inc. v. Massey* in 1993. In *Massey,* the Environmental Defense Fund (EDF) sought a declaration that the National Science Foundation had violated NEPA by going ahead with plans to incinerate food wastes in Antarctica without first having filed an environmental impact statement. The EDF also sought injunctive relief (that is, a court order) prohibiting further incineration. The D.C. Circuit described three exceptions to the presumption against extraterritoriality: the clear intent exception, the adverse effects exception, and the location of conduct exception. The first of these has already been described above. The second applies "where the failure to extend the scope of the statute to a foreign setting will result in adverse effects within the United States." And the third applies where the "conduct regulated by the government occurs within the United States." This would apply to federal agency decisions taken within the United States, even if the effects were largely felt outside the United States.

In concluding that the incineration project required an environmental impact statement, the Court of Appeals was influenced by the consideration that in Antarctica there is no foreign sovereign whose sovereignty might be infringed upon by extraterritorial application of NEPA; it is thus not possible to say with certainty whether the same result might be reached in cases involving U.S. agency decisions with effects within the territories of other sovereigns. Some commentators have seen this as a fourth exception to the presumption against extraterritoriality: U.S. law may be applied extraterritorially where no other sovereign's law exists.

The Court of Appeals' holding would seem to require environmental impact statements for nearly all federal government actions overseas, though, as nearly all agency decisions are made in the United States. In practice, the applicability of this holding may be limited to situations involving areas beyond national jurisdiction, where no other law exists. As there is a well-developed body of maritime law, as a practical matter the holding may apply only to Antarctica and perhaps outer space.

In the wake of *Massey* two additional cases have addressed the extraterritorial application of NEPA. In *NEPA Coalition of Japan v. Aspin*, the federal district (not circuit) court for the District of Columbia held the Defense Department was not required to prepare an environmental impact statement for activities at U.S. military installations in Japan. (A federal district court is a trial court; a federal circuit court is an appellate court. Appeals from the D.C. district court are heard by the D.C. circuit court; the decisions of the D.C. circuit court create binding precedent for the D.C. district court, although not for other district courts. Confusingly, some state trial courts are called "circuit courts," and some state appellate courts are called "district courts.") The district court applied the clear intent test, finding none in the language of NEPA, and also distinguished the case from *Massey*, on the grounds that Antarctica was not a foreign country and Japan was.

In *Hirt v. Richardson*, on the other hand, the federal district court for the Western District of Michigan found that NEPA was at least potentially applicable to the shipment of plutonium to Canada. The court took into account the significant involvement of the U.S. Department of Energy in the transfer and the transboundary effects of an accident during the shipping process. But it declined to issue the injunction because of "weighty considerations of U.S. foreign policy, nuclear non-proliferation, and the general interests of the Executive in carrying out U.S. foreign policy[.]" In other words, the political question doctrine deprived the court of jurisdiction.

In addition, environmental aspects of the activities of the U.S. military outside the territory of the United States are governed by Executive Order 12,114, which requires the military to comply with the spirit and intent of NEPA when operating overseas. Military activities that will have significant environmental effects require environmental analysis and documentation, although NEPA environmental impact statements are not required.

NEPA is not the only U.S. environmental statute with at least potential extraterritorial application. One U.S. environmental statute with extraterritorial application, albeit aimed at conduct occurring within the United States, is the Clean Air Act, section 7415 of which deals with "[e]ndangerment of public health or welfare in foreign countries from pollution emitted in the United States." In addition to NEPA, other environmental statutes whose extraterritoriality has been addressed, directly or tangentially, by U.S. courts include the Comprehensive Environmental Resource Control and Liability Act (CERCLA), the Marine Mammal Protection Act (MMPA), and the Resource Conservation and Recovery Act (RCRA).

The Comprehensive Environmental Resource Control and Liability Act (CERCLA)

CERCLA, the Superfund Act, seems by its terms to be extraterritorially applicable. Section 9611 permits suits by foreign parties for the release of a hazardous substance "in or on the territorial sea or adjacent shoreline of a foreign country of which the claimant is a resident," on a basis of reciprocity. (Reciprocity requires that, if the situation were reversed, a comparable remedy would be available to the U.S. plaintiff in the foreign country.) In *United States v. Ivey,* the federal District Court for the Eastern District of Michigan considered the extraterritoriality not of CERCLA itself, but of a subpoena issued pursuant to it. In *Ivey* the Environmental Protection Agency (EPA) sought to recover the cost of cleaning up a Superfund site from the alleged owner of the site, a Canadian citizen living in Canada. The court agreed with the defendant that service of process on the defendant in Canada "was not valid because the nationwide service of process provisions of CERCLA did not extend to Canada," although it found that it had jurisdiction over the defendant on other grounds.

Marine Mammal Protection Act (MMPA)

The MMPA limits the number of dolphins that may be killed in U.S. commercial tuna-fishing operations. As a practical matter, it made tuna fishing commercially nonviable in the United States and initiated the long-running dolphin-tuna controversy between the United States and other countries, principally Mexico. (The MMPA also requires the United States to place a trade embargo against any country that does not "demonstrate it has a [comparable] regulatory program governing the taking of marine mammals.")

The MMPA led to a ban on the import of tuna from Mexico and other countries, and embroiled the United States in dispute resolution proceedings before a GATT dispute resolution panel. (GATT, the General Agreement on Tariffs and Trade, was the precursor to today's World Trade Organization, or WTO). The panel found in favor of Mexico, reasoning that parties to GATT could apply trade-related laws extraterritorially only as specifically addressed in the GATT agreement or if necessary to protect human life or health (Strom, 1995; Weiss, 1995). The panel's finding provoked outrage from environmentalists and set the stage for the ongoing tension between environmentalists and the WTO. This is often seen, incorrectly, as an inevitable result of an underlying tension between trade and the environment. In fact, as can be seen in the case of the Florida Everglades (discussed later in this chapter), free trade is often more environmentally beneficial than protectionism.

The Resource Conservation and Recovery Act (RCRA)

RCRA empowers the EPA to regulate, among other things, the import and export of hazardous wastes. RCRA was intended to regulate hazardous waste from generation to disposal, but in *Amlon Metals, Inc. v. FMC* the federal district court for the Southern District of New York declined to apply RCRA extraterritorially. The plaintiff, Amlon Metals, was a UK recycler of industrial wastes. Amlon Metals received twenty containers of waste heavily contaminated with the toxic chemical xylene from FMC, a U.S. pesticide company, allegedly in violation of its agreement with FMC. Amlon first sued FMC in the U.K. Commercial Court, which dismissed the suit on the grounds that "all the actions claimed to be taken by FMC took place in the United States and U.S. law would apply." Amlon then sued in a U.S. court (the Southern District of New York), claiming violation of various federal statutes, including RCRA. The court applied both the clear intent and the legislative history prongs of *Foley's* three-pronged test and concluded that Congress had not intended the citizen-suit provision of RCRA (upon which Amlon relied) to be applied extraterritorially. Amlon was thus left without recourse for the RCRA claims: neither the UK court nor the U.S. court would provide a forum.

A foreign plaintiff wishing to sue in the courts of the United States for environmental harm committed by a U.S. juridical person (that is, either a natural person or some legal entity such as a corporation) abroad faces formidable practical obstacles. Even where a U.S. law has extraterritorial application or a rule of international law creates rights enforceable by individuals in U.S. courts, there are barriers of distance, language, and expense to overcome. Once the plaintiff makes it through the (metaphorical) courthouse door, she may find the defendant moving to dismiss on the grounds of *forum non conveniens*—that is, because the defendant thinks the U.S. court is an inconvenient forum for the dispute, and some other court, probably in the jurisdiction where the events occurred, would be more convenient. When this argument is made by a large corporation defending against a suit brought by an individual, it is made disingenuously: It is equally easy for Union Carbide to litigate in India or in the United States, but the amount of damages awarded by courts in India will be far smaller than the amount awarded by courts in the United States would be.

Even when jurisdiction and venue are established, the foreign plaintiff faces a difficult battle. The long-running litigation between Tom Beanal and Freeport-McMoRan, Inc., demonstrates the jurisdictional and practical difficulties presented in bringing a suit against a U.S. defendant in U.S. courts even for a plaintiff who is determined, patient, and backed by well-funded U.S. environmentalists. Tom Beanal is the leader of the Amungme Tribal Council; the Amungme are an indigenous people of the province of Indonesia formerly known as Irian Jaya (West Irian) and now known as West Papua, on the island of New Guinea. Freeport McMoRan operates the gigantic Grasberg copper, gold, and silver open-pit mine in West Papua. The Grasberg mine is the world's largest functioning open-pit mine. The opinion of the Court of Appeals for the Fifth Circuit states that the mine covers 26,400 square kilometers, which would make it larger than New Jersey and Delaware combined. The Amungme people live in this area, and their way of life has been adversely affected by the mine and the environmental damage accompanying it. They have also been subject to violence and exploitation on the part of mine employees and the Indonesian army, which has a presence in the area largely or entirely because of the mine.

The Fifth Circuit's 1999 opinion, the latest of many in the case, upheld a lower court's dismissal of Beanal's complaints, stating that "the sources of international law cited by Beanal . . . merely refer to a general sense of environmental responsibility and state abstract rights and liberties devoid of articulable or discernable standards and regulations to identify practices that constitute international environmental abuses or torts." In other words, the Fifth Circuit found no international agreement or rule of customary international law that created a right, enforceable by a foreigner in U.S. courts, to be free of environmental harm of the type alleged by Beanal. Any such rights, if they existed, could only come from U.S. law. And even in applying U.S. law, "federal courts should exercise extreme caution when adjudicating environmental claims under international law to ensure that environmental policies of the United States do not displace environmental policies of other governments. Furthermore, the argument to abstain from interfering in a sovereign's environmental practices carries persuasive force especially when the alleged environmental torts and abuses occur within the sovereign's borders and do not affect neighboring countries." The court was unwilling to apply U.S. environmental law extraterritorially, even to the actions of a U.S. party, where the actions took place within the territory of a foreign sovereign. In its unwillingness the court preserved the fiction that the environmental laws of Indonesia are likely to be enforced to the benefit of the Amungme people, even though the Indonesian government has a decades-long record of mistreatment of the inhabitants of West Irian and much of the province is or has been in open rebellion against the central government.

The *Beanal* case illustrates the extremely limited options available to a foreign party seeking redress in U.S. courts for environmental wrongs committed by U.S. parties outside the United States. The situation is made even more difficult when the plaintiffs are, like Tom Beanal, members of indigenous groups that are at odds with their own national government. But *Beanal* also illustrates another development: the integration of at least some areas of environmental law with human rights law and the emergence of a concept of green rights. Beanal is not primarily an environmental case; it is a human rights case. Tom Beanal's claims are human rights claims, brought under U.S. statutes such as the Alien Tort Claims Act and the Torture Victim Protection Act. In the complaints and the plaintiff's briefs, the environmental claims are seamlessly interwoven with the claims of torture, murder, and other human rights abuses. The court separated these claims from the others with its statement that breach of a "general sense of environmental responsibility" creates no rights enforceable under the Alien Tort Claims Act. It did not question that more traditional human rights violations—torture and murder—create enforceable rights under the act, although it dismissed those portions of the complaint on other grounds (lack of specificity in the pleadings).

The Environment and Trade: The Case of the Everglades

As we saw in Chapter 2, barriers to trade are a major cause of southern resentment of the North; many in the South quite correctly blame trade barriers for the continuing poverty of much of the South. In some cases protectionist trade barriers are directly linked not only to global poverty, but also to environmental harm.

The GATT/WTO treaty regime that governs international trade has been under steady attack by organized protest movements since the "Battle of Seattle" in 1999, at which protesters successfully disrupted WTO trade talks. While a large number of protesters seem to be motivated by simple alienation or anticorporate sentiment, two distinct ideological strands are present. Organized labor objects to the lowering of trade barriers that protect U.S. industries, because of the potential loss of U.S. jobs to overseas manufacturers. And environmentalists fear that U.S. companies will take advantage of laxer environmental standards in other countries or that they will be displaced by other companies that do. (The same concerns are equally present in other developed countries, of course; European and Japanese protesters worry about the loss of European and Japanese jobs and goods produced in environmentally unsound ways entering Europe and Japan.)

Both of these ideological positions are ill considered. While labor protectionism may benefit workers in the short run, in the long run it will prove disastrous. A free market in labor may result in a gradual loss of jobs to overseas competitors, giving the United States time to develop new industries and create new jobs, as it has done so often in the past. A protected labor market will result in increasingly uncompetitive industries; rather than being lost bit by bit, jobs will be lost all at once as these uncompetitive industries finally collapse, catastrophically.

The environmental objections to free trade may be equally ill considered. While it is not at all clear that increasing wealth in developing nations leads to environmental harm, it is clear that in at least some instances protectionism in developed countries does. The Florida Everglades, a United Nations Educational, Scientific, and Cultural Organization (UNESCO) World Heritage Site, is being destroyed by the South Florida sugar industry. In a world without trade barriers, however, no one would ever have bothered to grow sugar in the Everglades; only a century of subsidies and protectionism has made the Florida sugar industry viable (Monahan, 1992; Roberts, 1999). As a result, consumers in the United States now pay nearly three times as much for sugar as consumers in the rest of the world. This artificial price is supported by a complex system of direct and indirect subsidies, tariffs, and non-tariff barriers such as quotas, producing a tremendous benefit for a small number of U.S. producers, a heavy but bearable burden for U.S. consumers, and an environmental catastrophe for South Florida and the Florida Bay.

Even after a century of damage, much of the Everglades could still be saved and much developing-country poverty eliminated, if the United States could find the political will to eliminate tariff and non-tariff barriers to sugar imports. In addition, the $2 billion or so in excess of market price that U.S. consumers presently spend on overpriced sugar could be saved or otherwise spent, possibly producing economic benefits.

In the early twentieth century, after initial attempts to settle and farm the Everglades had failed disastrously, the federal government spent billions of dollars building an elaborate and ecologically disastrous hydraulic system in South Florida, the major components of which were the Hoover Dike around the southern shore of Lake Okeechobee and a network of drainage canals to lower water levels. This drainage has led to saltwater intrusion, soil fires, and soil subsidence. Research subsidies led to the development of special strains of sugarcane that could thrive in the (relatively) cooler climate of South Florida, but sugarcane needs nutrients not found in the soil of the Everglades. Runoff of agricultural chemicals, especially fertilizers containing phosphorus, is another leading cause of environmental damage (McCally, 1999).

The draining of the northern Everglades alone was insufficient to enable the Florida sugar industry to exist and prosper, though. The federal government, first through a U.S. Employment Service program so harsh that it eventually led to the indictment of a sugar company for peonage and later through a migrant worker program that also led to human rights complaints, helped to reduce sugar companies' labor costs (McCally, 1999). Most of all, barriers to trade kept sugar from other countries from competing in U.S. markets. Other sugarcane-growing countries—Brazil, India, Cuba, Mexico, the Dominican Republic—have climates and soil more suitable for sugar production and lower labor costs. The United States excludes sugar produced in these and other countries (including developed-world sugar producers such as Australia, Canada, and the European Union, the latter two of which grow sugar beets rather than sugarcane) through a two-part trade barrier system. The first component of the system is a loan program that guarantees a support price of eighteen cents per pound for cane sugar produced in the United States; the second is a tariff/quota regime that limits the amount of sugar that may enter the United States from abroad and heavily taxes the small amount that does.

The dispute over sugar tariffs and quotas has already found partial resolution in a trade rather than environmental context. The WTO and NAFTA have forced the United States to weaken its protectionist regime slightly, especially with regard to Canada and Mexico. Mexico, which heavily subsidizes its own inefficient sugar industry, has also made concessions. But in the case of the Everglades, at least, globalization of trade is the ecosystem's only hope; demand for sugar is unlikely to decrease, but sugar production can be shifted to regions less environmentally sensitive than the Everglades.

The Environment and War

All countries may at times find themselves at war, but warfare is a particular problem for the United States. The United States maintains the world's largest military. Its forces are continuously present outside the borders of the United States and frequently involved in conflict. This section examines two international environmental disputes arising from the conduct of war by the United States. The first, arising from the destruction of a chemical factory and storage complex at Pančevo, Serbia, during the Kosovo War, is particularly instructive because it employs three distinct dispute-resolution tools: an action before the ICJ, a UNEP fact-finding mission, and an investigation by the Office of the Prosecutor (OTP) of the International Criminal Tribunal for the Former Yugoslavia (ICTY). The second is an example, like the anti-globalization cause or the controversy over bio-prospecting, of misplaced zeal: the dispute over U.S. use of munitions and shielding made from depleted uranium.

The law of war can be divided into two parts: the *jus ad bellum*, or law pertaining to the use of force, and the *jus in bello*, or law pertaining to the conduct of war. The Pančevo and depleted uranium controversies both involve questions of the *jus in bello*. The law of environmental damage during the conduct of war has a long history. War has been an unwelcome but constant factor in human history; and as long as there has been war, there has been environmental damage, both incidental and deliberate, resulting from it. More than 2,000 years ago the Romans sowed the soil of defeated Carthage with salt, and they were not the first to make use of environmental warfare.

In modern times regulation of instrumentalities of war with a potential impact on the environment dates at least from the 1868 Declaration of St. Petersburg, which stated "the only legitimate object which states should endeavor to accomplish during war is to weaken the military forces of the enemy." This, of course, leaves open the possibility of direct attacks on the environment, such as Iraq's release of oil into the Persian Gulf during the first Gulf War or the United States' use of Agent Orange to destroy forests during the Vietnam War, if the only purpose of the attacks is to weaken the military forces of the enemy, rather than to harm the civilian population.

A more comprehensive attempt to codify the laws of war took place at the end of the nineteenth century and the beginning of the twentieth century. The 1899 and 1907 Hague Conventions provided that "the right of belligerents to adopt means of injuring the enemy is not unlimited," and provided for the protection of property (theoretically, at least, including environmental resources).

The Hague Conventions, however, failed to prevent the use of poison gas in World War I; after the war (in 1925), the Geneva Gas Protocol was adopted to address this specific problem and for the most part seems to have been effective. The goal of the Gas Protocol, of course, is the protection of human beings, not the environment, from poison. The environmental law of war, and the idea of human environmental rights, would not be significantly developed until after World War II.

Pančevo

During the war between Serbia-Montenegro (then generally known as the Federal Republic of Yugoslavia) and the North Atlantic Treaty Organization (NATO), U.S. bombers under NATO command repeatedly attacked a petrochemical and fertilizer factory complex at Pančevo, in Serbia. The Pančevo complex was selected as a target because it was believed by NATO to produce military chemicals as well as civilian ones. Pančevo's mayor reported that the complex was struck by at least fifty-six aircraft-launched missiles on twenty-three days between March 24 and June 8, 1999. The complex's chemical storage tanks, along with production facilities, were destroyed; thousands of tons of toxic chemicals were released to run into the Danube (Booth, 1999; Schmetzer, 1999). (It should be pointed out here that this shows poor planning on the part of the complex's designers. Even in the absence of war, the storage tanks might have been damaged or destroyed by fire, earthquake, or accident; a drainage system and catch basin should have been built so that the runoff from such an event would not end up in the Danube.)

In addition to the damage to the Danube, the air in Pančevo was filled with fumes for several days, causing respiratory and stomach ailments. Leaves turned yellow or black; fish caught in the Danube looked sickly, and local government officials temporarily banned fishing.

As a peacetime incident the pollution from Pančevo might have been viewed as catastrophic; in the larger context of the war it attracted less attention. Nonetheless, in the new climate of environmental awareness, UNEP dispatched a fact-finding team to Yugoslavia. The team, led by Pekka Haavisto, found no evidence of an environmental "disaster," but concluded that measurable damage had occurred. (Environmental management in Serbia was generally poor; the UNEP report noted that some or even most of the environmental damage it found might have predated NATO's attacks.) Pekka Haavisto reported that there was no "major ecocide or countrywide catastrophe," but rather a few "chosen hot spots where immediate action has to take place." Of these hot spots, the hottest was Pančevo (UNEP, 1999).

UNEP's investigation was one of three mechanisms employed to resolve the environmental dispute between Serbia-Montenegro and the NATO nations resulting from the destruction of Pančevo. The UNEP fact-finding mission looked at the environmental aspects of the problem alone and was concerned with questions of fact rather than questions of law. The other mechanisms considered the environmental damage at Pančevo as a relatively small event in the context of the larger war.

While the war was still going on, the Federal Republic of Yugoslavia brought an action before the ICJ against ten NATO nations involved in the war: Belgium, Canada, France, Germany, Italy, the Netherlands, Portugal, Spain, the United Kingdom, and the United States. Cases against two parties (Spain and the United States) were quickly dismissed on jurisdictional grounds; the ICJ declined to order a halt to the conflict, but the other eight cases remained before the court until 2004, when they too were dismissed for lack of jurisdiction. Among the illegal harms alleged in the lawsuits was the environmental damage from the destruction of the complex at Pančevo.

Also while the war was still in progress, the Office of the Prosecutor of the International Criminal Tribunal for the former Yugoslavia established a committee to report on war crimes allegedly committed by NATO during the conduct of the war, including the destruction of the chemical complex at Pančevo. A year later, the OTP's Final Report, issued on June 13, 2000, stated that none of NATO's actions merited further investigation or prosecution before the ICTY.

A prosecutor's report has no precedential value even in a common-law system such as that of the United States, let alone before an international tribunal. As a practical matter, however, the OTP's report may effectively settle the question: The destruction of the complex at Pančevo was not illegal under international environmental law. It seems unlikely that the ICJ, should it ever work its way around to the question of Pančevo, will reach a different conclusion.

In reaching its own conclusion, the OTP made some surprising observations about the law of war and the environment. The fundamental rule in cases such as these, according to the OTP, was set forth in Article 55 of Protocol I to the Geneva conventions:

> Care shall be taken in warfare to protect the natural environment against widespread, long-term and severe damage. This protection includes a prohibition of the use of methods or means of warfare which are intended or may be expected to cause such damage to the natural environment and thereby to prejudice the health or survival of the population.

While recognizing that the United States was not a party to Protocol I, the OTP considered the possibility that Protocol I in general, and particularly Article 55, might represent accurate statements of obligations existing under customary international law. Although this is not at all clearly established in international law generally, it is for the most part consistent with U.S. practice.

While the idea that Article 55 might state a rule of customary international law is encouraging from an environmentalist perspective, the OTP's extremely high standard set by the rule is less so: "[W]idespread, long-term, and severe . . . is a triple, cumulative standard" that is not met by "ordinary battlefield damage of the kind caused to France in World War I." Rather, the harm must "be measured in years rather than months." The use of the World War I example is especially disheartening; both humanity's understanding of the environment and the destructive potential of war have increased enormously since then, and there has been a fundamental shift in environmental values.

The OTP theorized that even the damage caused by oil spills and fires in the first Gulf War might not meet the Protocol I standard, although the harm was far greater than that at Pančevo and the military necessity probably nonexistent. The OTP could easily have found that the Pančevo bombing was not a prosecutable environmental war crime even under a post-Gulf War I standard. (The evidentiary problem alone—the difficulty or impossibility of distinguishing environmental damage caused by NATO's actions from preexisting damage caused by careless management—would have sufficed.) Instead, its report leaves the reader wondering what actions in twentieth or twenty-first century warfare, if any, might constitute environmental war crimes.

Depleted Uranium

During the Kosovo war, and more recently during the Iraq war, the United States has been criticized for its use of munitions made from depleted uranium. These criticisms are based on a fundamentally incorrect understanding of the facts; as with biopiracy and foreign trade, activists have once again picked the wrong battle. This is symptomatic of a problem that affects both international environmental law and international environmental activism and is particularly evident in attempts to bring about environmentally sound rules in two areas: the law of international trade and the law of war. Opponents of "globalization" lack a clear focus, but opponents of war do not.

No one likes war; and an unfortunate truth for citizens of the United States is that few or no wars in recent U.S. history can properly be characterized as defensive. To many antiwar activists, who often make common cause with, and in fact are often the same individuals as, environmental activists, "war" has unfortunately become synonymous with "U.S. foreign policy"; hatred for one leads to hatred of the other. This in turn leads to an unwillingness to compromise or to acknowledge scientific evidence that certain military activities may not be all that environmentally harmful. Opposition to depleted uranium, a metal used to make munitions and armor plating, is an example of this problem.

The phrase "depleted uranium" is misleading and probably the primary cause of the popular misconceptions surrounding the substance. The name conjures up images of nuclear waste left over from power plants being fired from antitank weapons and scattered across the landscape of Kosovo, Iraq, or any place where the U.S. military is active. Part of the reaction to depleted uranium stems from a misunderstanding of its nature. Depleted uranium is not material that has been used in nuclear power plants; it is uranium from which the radioactive isotopes have been extracted, leaving an inert metal heavier than lead. There is some residual radioactivity, because the extraction process isn't perfect, but not much. Naturally occurring uranium is composed mostly of two isotopes: Uranium–235 (^{235}U) and Uranium–238 (^{238}U), in a ratio of about 0.7 percent ^{235}U to 99.3 percent ^{238}U. Nuclear power plants use ^{235}U, the radioactive isotope. The extraction of this isotope leaves a large amount of nonradioactive ^{238}U. Uranium–238 is an unusually dense substance, much more so than lead. Thus it is suited for making military projectiles: More kinetic energy can be packed into a smaller object (Royal Society, 2001).

Uranium is chemically poisonous, of course, like many metals; ingesting or inhaling it can be harmful just as ingesting or inhaling lead might be (Royal Society, 2002). Its use may have military advantages, but those may be outweighed by the public relations problems that are attached to it. (Even the use of nuclear weapons may not actually violate international law, according to an advisory opinion of the ICJ. But there are sound reasons for avoiding their use, and countries with nuclear weapons do so despite the tactical advantages that might be achieved.) And depleted uranium is only economically feasible to use because it is available as a by-product of the nuclear power and weapons industry; in the absence of an industry that produced ^{238}U as a by-product of the more valuable ^{235}U, ^{238}U would probably be too expensive to use for weapons. Uranium is only about 68 percent heavier than an equivalent volume of lead; a weapon firing depleted uranium rounds of the same volume and at the same velocity as lead rounds has about 68 percent more punch. (Of course, more propellant must be used to achieve this, and the same result could probably be obtained by using longer lead projectiles, or by increasing the caliber of the weapon, or by increasing the velocity of the projectile, or by firing 68 percent more rounds. The "self-sharpening" properties of uranium rounds also make them ideal for armor-piercing, but alternatives are available.) Tungsten, the metal used to make light bulb filaments, is even heavier than uranium, and far less toxic; it has been proposed as a replacement for toxic lead munitions (*Economist*, 1999). Similar concerns have led many states to ban lead shot for waterfowl hunting: Steel shot is more expensive but does not pollute the water into which it falls.

But the opposition to depleted uranium weapons does not focus on these entirely rational arguments, but on scientifically unfounded worries about radiation. This argument serves only to discredit those who make it.

Is a Fifth Geneva Convention Needed?

Some NGOs, including Greenpeace, have expressed the belief that the existing law relating to the conduct of war is inadequate to protect the environment during armed conflicts and have called for a fifth Geneva Convention. The Legal Committee of the United Nations General Assembly has discussed a convention aimed specifically at the protection of the environment during war; the United States opposes the idea on the grounds that existing international law, if complied with, already provides adequate protection.

Summary

This discussion of special issues for the United States in international environmental law addresses the relationship between the U.S. legal system and the international legal system. Self-executing international agreements have a status equivalent to federal statutory law in the United States: Provisions of self-executing treaties may control over earlier inconsistent statutory provisions, while later inconsistent statutory provisions will control even over a provision in a self-executing treaty. Non-self-executing treaties create international obligations but are not enforceable in U.S. courts unless Congress has passed implementing legislation. U.S. courts may apply customary international law where there exists no U.S. law on point; when there is conflicting U.S. law on point, the applicability of customary international law is unclear. International agreements, as federal law in an area delegated to the federal government by the Constitution (foreign affairs), preempt state law.

The United States has been reluctant to apply its own environmental laws to activities by U.S. citizens outside the United States. NEPA has been applied to activities in Antarctica but not to activities within the territories of other sovereign states.

Like most developed countries, the United States protects its farmers by restricting market access for foreign agricultural products, to the detriment of farmers in the South. In the case of the Florida Everglades, this agricultural protectionism is harmful to a delicate and unique ecosystem.

The constant presence of U.S. military forces outside the United States, together with the frequent involvement of the United States in armed conflict, creates the potential for many international environmental disputes. One of these disputes, over depleted uranium, is without foundation. But the more serious dispute, the bombing of Pančevo, raises important questions about the environmental responsibility of states during the conduct of war. In this particular case, a UNEP task force has found that the damage was less serious than initially feared and not necessarily or entirely the fault of the United States. The Office of the Prosecutor of the International Criminal Tribunal for the Former Yugoslavia has found no grounds for going forward with a criminal prosecution against U.S. forces involved in the attack on Pančevo, and the International Court of Justice has dismissed Serbia-Montenegro's case against the United States for, among other things, the Pančevo bombing.

Sources and Further Reading
Books and Articles

American Law Institute. *Restatement (Third) of the Foreign Relations Law of the United States*. Philadelphia: ALI, 1987.

Baker, Betsy. *Legal Protections for the Environment in Times of Armed Conflict*. 33 Virginia Journal of International Law 351 (1993).

Bederman, David J. *International Law Frameworks*. New York: Foundation Press, 2001.

Booth, William. *A Ghost City of Mixed Poisons—NATO Bombs Left Site of Petrochemical Complex a Toxic Slough*. Washington Post, A-15. 21 July 1999.

Brass Hats Lead to Tungsten. The Economist, 31 July 1999.

Buergenthal, Thomas, and Sean D. Murphy. *Public International Law in a Nutshell, 3rd ed.* St. Paul, MN: West, 2002.

Danish Institute for International Affairs (Dansk Udenrigspolitisk Institut). *Humanitarian Intervention: Legal and Political Aspects*, 1999.

Douglas, Marjorie Stoneman. *Everglades: River of Grass*. New York: Rinehart and Company, 1947.

Drumbl, Mark. *Waging War Against the World: The Need to Move from War Crimes to Environmental Crimes*. 22 Fordham International Law Journal 122 (1998).

Fair, Karen. *Environmental Compliance in Contingency Operations: In Search of a Standard?* 157 Military Law Review 112 (1998).

Kibel, Paul Stanton. *The Earth on Trial: Environmental Law on the International Stage*. New York: Routledge, 1999.

Mahler, Vincent A. *The Political Economy of North-South Commodity Bargaining: The Case of the International Sugar Agreement*. 38 International Organizations 709 (1984).

McCally, David. *The Everglades: An Environmental History*. Gainesville: University Press of Florida, 1999.

Monahan, Katherine E. *U.S. Sugar Policy: Domestic and International Repercussions of Sour Law*. 15 Hastings International and Comparative Law Review 325 (1992).

Morris, Virginia. *Protection of the Environment in Wartime: The United Nations General Assembly Considers the Need for a New Convention*. 27 International Lawyer 775 (1993).

Okorodudu-Fubara, Margaret T. *Oil in the Persian Gulf: Legal Appraisal of an Environmental Warfare*. 23 St. Mary's Law Journal 123 (1991).

Riechel, Sylvia M. *Governmental Hypocrisy and the Extraterritorial Application of NEPA*. 26 Case Western Reserve Journal of International Law 115 (1994).

Roberts, Adam and Richard Guelff, eds. *Documents on the Law of War, 2nd ed.* Oxford: Oxford University Press, 2000.

Roberts, Paul. *The Sweet Hereafter: Our Craving for Sugar Starves the Everglades and Fattens Politicians*. Harper's, 1 November 1999.

The Royal Society. *The Health Hazards of Depleted Uranium Munitions, Part I,* May 2001. Available at http://www.royalsoc.ac.uk/policy/ (visited September 26, 2004).

———. *The Health Hazards of Depleted Uranium Munitions, Part II,* March 2002. Available at http://www.royalsoc.ac.uk/policy/ (visited September 26, 2004).

Sands, Philippe (moderator). *The Gulf War: Environment As a Weapon.* 85 American Society of International Law Proceedings 214 (1991).

Schmetzer, Uli. *Serbs Allege NATO Raids Caused Toxic Catastrophe: Bombed Refineries, Plants Spewed Stew of Poisons, They Say.* Chicago Tribune, 8 July 1999.

Schmitt, Michael N. *Green War: An Assessment of the Environmental Law of International Armed Conflict.* 22 Yale Journal of International Law 1 (1997).

Schwabach, Aaron. *Environmental Damage Resulting From the NATO Military Action Against Yugoslavia.* 25 Columbia Journal of Environmental Law 117 (2000).

———. *How Free Trade Can Save the Everglades.* 14 Georgetown International Environmental Law Review 301 (2001).

———. *NATO's War in Kosovo and the Final Report to the Prosecutor of the International Criminal Tribunal for the Former Yugoslavia.* 9 Tulane Journal of International and Comparative Law 167 (2001).

Sharp, Major Walter G. Sr. *The Effective Deterrence of Environmental Damage During Armed Conflict: A Case Analysis of the Persian Gulf War.* 137 Military Law Review 1 (1992).

Small, John Kunkel. *From Eden to Sahara: Florida's Tragedy.* Lancaster, PA: Science Press Printing, 1929.

Smith, Ian. *Prospects for a New International Sugar Agreement.* 17 Journal World Trade Law 308 (1983).

———. *UNCTAD. Failure of the UN Sugar Conference.* 19 Journal of World Trade Law 296 (1985).

The Spoils of War: What Can the Past Tell About the Effect of Military Conflict on the Environment? The Economist, 27 March 2003.

Strom, Torsten H. *Another Kick at the Can: Tuna/Dolphin II.* 33 Canadian Year Book of International Law 149 (1995).

Taylor, Jennifer Dawn. *Ambiguities in the Foreign Corrupt Practices Act.* 61 Louisiana Law Review 861 (2001).

Trimble, Phillip R. *International Law: United States Foreign Relations Law.* New York: Foundation Press, 2002.

Weiss, Friedl. *The Second Tuna GATT Panel Report.* 8 Leiden Journal of International Law 135 (1995).

Cases, Statutes, and Other U.S. Legal Materials

American Banana Co. v. United Fruit Co., 213 U.S. 347 (1909).

Amlon Metals, Inc. v. FMC, 775 F. Supp. 668 (S.D.N.Y. 1991).

Banco Nacional de Cuba v. Sabbatino, 376 U.S. 398 (1964).

Beanal v. Freeport-McMoRan, Inc., 969 F. Supp. 362 (E.D. La. 1997), aff'd 197 F.3d 161 (5th Cir. 1999).

Comprehensive Environmental Resource Control and Liability Act (CERCLA), 26 U.S.C. § 4611–4682; major amendments in 1983 and 1986, reconstituted at 26 U.S.C. § 9507 and 42 U.S.C. § 9601–57.

Constitution of the United States.

Environmental Defense Fund, Inc. v. Massey, 986 F.2d 528 (D.C. Cir. 1993).

Environmental Effects Abroad of Major Federal Actions, Executive Order 12,114, 44 Fed. Reg. 1957 (1979), 42 U.S.C. § 4321 (1982).

Equal Opportunity Employment Commission v. Arabian American Oil Co., 499 U.S. 244 (1991).

Federal Agriculture Improvement and Reform Act of 1996, § 156, Pub. Law 104–127, 4 April 1996, 110 Stat. 888.

Florida Sugar Cane League, Inc. v. South Florida Water Management District, 617 So. 2d 1065 (Fla. Dist. Ct. App. 1993).

Foley Bros., Inc. v. Filardo, 336 U.S. 281 (1949).

Foreign Corrupt Practices Act, 15 U.S.C. § 78m(b), 15 U.S.C. § 78dd–1 (1988) and 15 U.S.C. § 78dd–2.

Foreign Sovereign Immunities Act of 1976, 28 U.S.C. §§ 1330, 1332(a), 1391(f) and 1601–11.

Heartland By-Products, Inc. v. United States, 74 F. Supp. 2d 1324, 1326–27 (Court of International Trade 1999).

Hirt v. Richardson, 127 F. Supp. 2d 849, (W.D. Mich. 2001).

Marine Mammal Protection Act (MMPA), 16 U.S.C. § 1361–1407.

Miccosukee Tribe of Indians of Florida v. United States Environmental Protection Agency, 105 F. 3d 599 (11th Cir. 1997).

Miccosukee Tribe of Indians v. Florida Department of Environmental Protection, 656 So. 2d 505 (Fla. Dist. Ct. App. 1995).

National Environmental Policy Act, 42 U.S.C. § 4321–4375.

NEPA Coalition of Japan v. Aspin, 837 F. Supp. 466 (D.D.C. 1993).

Resource Conservation and Recovery Act (RCRA), 42 U.S.C. § 6901–6992.

The Paquete Habana, 175 U.S. 677 (1900).

United States Cane Sugar Refiners' Association v. Block, 544 F. Supp 883 (Court of International Trade 1982), aff'd, 683 F. 2d 399 (Court of Customs and Patent Appeals 1982).

United States Customs Service, *What Every Member of the Trade Community Should Know About Cane and Beet Sugar: Quota, Classification and Entry* 13 (2000).

United States v. Ivey, 747 F. Supp. 1235 (E.D. Mich. 1990); related Canadian decision, *United States v. Ivey* [1995], 26 O.R. 3d 533 (Ont. Gen. Div.), affirmed [1996], 30 O.R. 3d 370 (Ont. Ct. App.).

United States v. South Florida Water Management District, 922 F. 2d 704 (11th Cir. 1991); cert. denied, 502 U.S. 953 (1991); on remand, 847 F. Supp. 1567 (S.D. Fla. 1992); aff'd in part, rev'd in part, 28 F. 3d 1563 (11th Cir. 1994); cert. denied 514 U.S. 1107 (1995).

Treaties and Related Documents

Convention for the Amelioration of the Condition of the Wounded and Sick in Armed Forces in the Field, Aug. 12, 1949, 6 U.S.T. 3114, 75 U.N.T.S. 31 [Geneva Convention I].

Convention for the Amelioration of the Condition of the Wounded, Sick and Shipwrecked Members of the Armed Forces at Sea, Aug. 12, 1949, 6 U.S.T. 3217, 75 U.N.T.S. 85 [Geneva Convention II].

Convention Relative to the Protection of Civilian Persons in Time of War, Aug. 12, 1949, 6 U.S.T. 3516, 75 U.N.T.S. 287 [Geneva Convention IV].

Convention on the Prohibition of Military or Any Other Hostile Use of Environmental Modification Techniques, Dec. 10, 1976, 31 U.S.T. 333, 1108 U.N.T.S. 151.

Convention Relative to the Treatment of Prisoners of War, Aug. 12, 1949, 6 U.S.T. 3316, 75 U.N.T.S. 135 [Geneva Convention III].

Convention Respecting the Laws and Customs of War on Land, July 29, 1899, Reprinted in James Brown Scott, The Hague Conventions and Declarations of 1899 and 1907, at 100 (1918) [1899 Hague Convention].

Convention Respecting the Laws and Customs of War on Land, Oct. 18, 1907, 36 Stat. 2277, 205 Consol. T.S. 277 [1907 Hague Convention]. The

Convention was the fourth of thirteen to emerge from the 1907 Hague Peace Conference and is thus known as "Hague IV."

Declaration Renouncing the Use, in Time of War, of Explosive Projectiles Under 400 Grammes Weight. Saint Petersburg, Nov. 29/Dec. 11 1868, 138 Consol. T.S. 297. Reprinted in The Laws of Armed Conflicts: A Collection of Conventions, Resolutions, and Other Documents 102–03 (Dietrich Schindler & Jiri Toman eds., 1988). The odd date (Nov. 29/Dec. 11) is the result of a difference between the Russian and Gregorian calendars that persisted until the time of the Russian Revolution.

Protocol Additional to the Geneva Conventions of 12 August 1949, and Relating to the Protection of Victims of International Armed Conflicts, June 8, 1977, 1125 U.N.T.S. 3 [Protocol I].

Protocol Additional to the Geneva Conventions of 12 August 1949, and Relating to the Protection of Victims of Non-International Armed Conflicts, June 8, 1977, 1125 U.N.T.S. 609 [Protocol II].

Protocol for the Prohibition of the Use in War of Asphyxiating, Poisonous, or Other Gases and of Bacteriological Methods of Warfare, June 17, 1925, 26 U.S.T. 571, T.I.A.S. No. 8061.

Other International Materials

Final Report to the Prosecutor by the Committee Established to Review the NATO Bombing Campaign Against the Federal Republic of Yugoslavia. Available at http://www.un.org/icty/pressreal/nato061300.htm (visited May 28, 2004).

Legality of the Threat or Use of Nuclear Weapons (advisory opinion). 1996 I.C.J. 226 (1996).

S.C. Res. 674, U.N. SCOR, 2951st mtg., *U.N. Doc. S/RES/674* (1990).

S.C. Res. 686, U.N. SCOR, 2978th mtg., *U.N. Doc. S/RES/686* (1991).

S.C. Res. 687, U.N. SCOR, 2981st mtg., *U.N. Doc. S/RES/687* (1991).

UNEP. *The Kosovo Conflict: Consequences for the Environment and Human Settlements* (1999). Available at http://www.grid.unep.ch/btf/final/index.html (visited May 28, 2004).

Other U.S. Materials

Environmental Effects Abroad of Major Federal Actions. Executive Order 12,114, 44 Fed. Reg. 1957 (1979), 42 U.S.C. § 4321 (1982).

International and Operational Law Department. *Operational Law Handbook.* The Judge Advocate General's School, U.S. Army, JA 442 (1996): 18–1, 18–2.

Policy Review Directive/NSC-23, Apr. 8, 1993. *United States Policy on Extraterritorial Application of the National Environmental Policy Act (NEPA).*

United States Department of the Army. *United States Environmental Strategy into the 21st Century* (1992).

4

Chronology

1625 Huig de Groot, today better known as Grotius, publishes *De Jure Belli ac Pacis* (*On the Law of War and Peace*) in Paris; the book comes to be seen as a foundational text of modern international law. The 1646 (7th) edition contains an ambivalent expression of transboundary resource law. On the one hand, there is absolute sovereignty over resources within a nation's territory: A river "is the property of the people through whose territory it flows, or the ruler under whose sway that people is . . . to them all things produced in the river belong." On the other hand, "the same river . . . has remained common property." The tension between territorial sovereignty and territorial integrity that gives rise to this ambivalence has remained a defining feature of international environmental law.

1661 John Evelyn publishes *Fumifugium*, in which he recommends that the burning of coal be prohibited in London so as to eliminate the "Hellish and dismall cloud of seacoale" that lies over the city. His advice is not immediately followed.

1815 The modern world's first international organization, the Danube Commission, is founded. Although primarily concerned with navigation, over the course of the nineteenth century the Danube Commission adopts a number of proto-environmental regulations regarding dumping in the river and transport of explosive and toxic materials.

135

1827 French mathematician Jean Baptiste Joseph Fourier becomes the first of several nineteenth-century scientists to discover the existence of the greenhouse effect: Fourier deduces that the sun's heat is partially trapped in the earth's atmosphere.

1852 Acid rain is reported in and around Manchester, England.

1873 Nearly 1,200 people die in London over a three-day period from severe air pollution mixed with fog. These "killer fogs" are to become a commonplace feature of London for the next eighty years.

The International Law Association is founded in Brussels, Belgium, for the "study, elucidation and advancement of international law and the furthering of international understanding and goodwill."

1885 Karl Benz builds the first successful gasoline-driven automobile.

1896 Swedish scientist Svante Arrhenius names the greenhouse effect. Arrhenius suggests that carbon dioxide is responsible for the greenhouse effect, that the amount of carbon dioxide in the atmosphere has been increasing since the beginning of the industrial revolution, and that, as a result, global temperatures may increase.

A lead and zinc smelting plant, built by a U. S. company, begins operating at Trail, British Columbia.

1906 The smelting plant at Trail is transferred to Canadian ownership.

1909 The United States and the United Kingdom (acting for Canada) sign the Boundary Waters Treaty, leading to the formation of the International Joint Commission. The commission prevents and resolves resource allocation and other environmental disputes between the United States of America and Canada arising from the use of boundary and transboundary waters. Its powers include reviewing, approving or disapproving, and regulating projects affecting those waters. Over the next century, its duties are expanded to include water and air quality.

1914   Martha, the last passenger pigeon, dies in the Cincinnati Zoo.

1915   Canada's Commission on Conservation sets forth an early expression of the idea of intergenerational equity: "Each generation is entitled to the interest on the natural capital, but the principal should be handed on unimpaired."

1916   Alarmed by the disappearance of the passenger pigeon and other great flocks of migrating birds, the United States and the United Kingdom (acting for Canada) sign the Migratory Bird Treaty to control hunting of migratory birds. The United States concludes a similar bilateral treaty, covering mammals as well as birds, with Mexico in 1936.

1925–   The smelter at Trail adds two 400-foot smokestacks,
1927   leading to an increase in the amount of sulfur dioxide from the smelter reaching the United States.

1927   In deciding the *Donauversinkung* water-resource dispute between two German states, the German high court holds that both states have rights in the shared resource. This balancing of the duty of states to do no harm to the territory of other states with the right of states to develop resources within their own boundaries is to become another fundamental principle of international environmental law.

1928   The United States and Canada submit the dispute over pollution from the smelter at Trail to the International Joint Commission established by the Boundary Waters Treaty of 1909.

1931   The International Joint Commission issues its report in the Trail Smelter dispute, recommending that Canada pay compensation to the United States and implement remedial measures to prevent further damage.

1935 The United States and Canada conclude a *compromis* convention; Canada agrees to pay $350,000 for damage caused by the Trail smelting plant up to 1932, and the two countries agree to submit the question of subsequent liability and prevention to an arbitral tribunal. The format of the arbitral tribunal is typical, with each side appointing an arbitrator and the two sides agreeing on a third, presumably neutral, arbitrator; the Trail Smelter arbitral tribunal thus has arbitrators from Canada, the United States, and Belgium.

1941 The Trail Smelter arbitral tribunal issues its landmark decision, setting forth in dicta the principle of state responsibility of transboundary environmental damage and ushering in the era of modern international environmental law. World War II and its aftermath delay further immediate development of this idea.

1945 World War II ends; the United Nations is founded, providing a world forum for discussion and advancement of international law.

1946 Two British military vessels strike mines off the coast of Albania in the Corfu Channel.

1948 The International Union for the Protection of Nature is founded; it will change its name to the International Union for Conservation of Nature and Natural Resources (IUCN) in 1956 and finally become IUCN-The World Conservation Union in 1990.

1949 Aldo Leopold publishes *A Sand County Almanac.*

The Diplomatic Conference for the Establishment of International Conventions for the Protection of Victims of War adopts the four Geneva Conventions.

In the *Corfu Channel Case* the International Court of Justice sets forth the principle that every state has the "obligation not to allow knowingly its territory to be used for acts contrary to the rights of other states."

The United Nations holds its first environmental conference, the Scientific Conference on the Conservation and Utilization of Resources, in Lake Success, NY.

1951   The multilateral International Plant Protection Convention is opened for signature.

1952   Severe air pollution mixed with fog kills 4,000 people in London in five days—the "Great Smog." Attention is focused on the problem of air pollution; in the ensuing years the United Kingdom will enact air pollution controls that will tremendously improve the quality of air in British cities. Some measures, such as the building of taller smokestacks, pose the possible expense of greater pollution—but little or no loss of human life—elsewhere, as pollution is carried farther from its source.

1953   Smog kills between 170 and 260 people in New York City over a six-day period.

1956   Dr. Hajime Hosokawa, an employee of the fertilizer, petrochemicals, and plastics company Chisso Corporation, links a mysterious outbreak of disease in the fishing and factory town of Minamata, Japan, to fish consumption. "Minamata disease" ultimately causes more than 100 deaths and hundreds of cases of brain damage, birth defects, and other severe illness in the fishing town of Minamata, Japan. Other investigators link the disease to contamination of fish by industrial waste dumped into Minamata Bay by the Chisso Corporation.

1957   The *Lac Lanoux* arbitral tribunal establishes limits to state responsibility: States may develop their natural resources without interference in cases in which no harm is done to neighboring states.

1958   In response to criticism, the Chisso Corporation stops dumping wastes containing mercury into Minamata Bay and begins dumping them into the Minamata River. Within months the inhabitants of the village of Hachimon on the Minamata River have begun to show symptoms of Minamata disease.

Oil is discovered in the Niger Delta in the British colony of Nigeria.

The first United Nations Conference on the Law of the Sea approves draft environmental protection provisions.

1959   Dr. Hosokawa conducts experiments identifying mercury in the bodies of fish as the cause of Minamata disease. His employer, Chisso Corporation, suppresses his results. Nonetheless other researchers reach the same conclusion in 1958 and 1959, and in 1959 villagers and environmentalists protest Chisso's continued dumping and demand compensation. Chisso reaches settlements with some villagers.

The Antarctic Treaty is signed.

1961   The Antarctic Treaty enters into force, setting aside Antarctica for scientific and peaceful purposes.

The WWF is founded as the World Wildlife Fund; later it becomes the Worldwide Fund for Nature and eventually "WWF—the Conservation Organization."

1962   Rachel Carson publishes *Silent Spring*, launching a nationwide, and eventually worldwide, debate about environmental policy and law.

1963   The United Kingdom, the United States, and the Soviet Union (but not France) sign the Treaty Banning Nuclear Weapon Tests in the Atmosphere, Outer Space and Under Water. Although opponents of nuclear weapons consider it a failure because it has the effect of legalizing underground nuclear tests and making them routine, it is an environmental triumph because it greatly reduces the far greater public health and environmental hazards from atmospheric nuclear tests. Continued testing of nuclear weapons by France in the South Pacific becomes a focus of protests by environmentalists and antinuclear activists.

The Convention Concerning the International Commission for the Protection of the Rhine Against Pollution establishes a joint environmental authority among the riparian states of the Rhine. The convention is superseded by the Convention for the Protection of the Rhine in 1999.

1967 The oil tanker *Torrey Canyon* spills over a 100,000 tons of oil into the sea near Great Britain, leading to demands for an improved international system for imposition of liability for oil spills.

In Vietnam, the U.S. military begins Operation Ranch Hand, a program to prevent the use of forests as cover by enemy forces. Operation Ranch Hand combines mechanical cutting of trees with the aerial spraying of chemical defoliants, leading to the loss of about one-seventh of South Vietnam's forests and half of its wetlands, and causing lasting health and environmental effects.

The Indonesian government grants the U.S. mining company Freeport-McMoRan mining rights to a large area in New Guinea.

The Treaty on Principles Governing the Activities of States in the Exploration and Use of Outer Space, Including the Moon and Other Celestial Bodies, better known as the Outer Space Treaty, is signed.

The International Law Association, a nongovernmental organization, promulgates its Helsinki Rules on the Uses of the Waters of International Rivers, a set of proposed rules for the management of transboundary freshwater resources.

1968 In Minamata, Japan, Chisso Corporation stops dumping wastes containing mercury.

The United Nations Educational, Scientific and Cultural Organization (UNESCO) organizes the Intergovernmental Conference of Experts on the Scientific Basis for Rational Use and Conservation of the Resources of the Biosphere. This is the first international conference of sovereign states to address the problem of reconciling the environment and development and is the precursor to later conferences on sustainable development.

The African Convention for the Conservation of Nature and Natural Resources is signed.

1969 The United States Congress enacts the National Environmental Policy Act (NEPA); President Nixon signs NEPA into law on January 1, 1970.

Two treaties on pollution from oil spills are signed: the International Convention Relating to Intervention on the High Seas in Cases of Oil Pollution Casualties and the International Convention on Civil Liability for Oil Pollution Damage.

1970 April 22, 1970, is the first Earth Day.

The United States Environmental Protection Agency, the first such agency in the world, is created.

Friends of the Earth International is founded.

1971 A small group of U.S. and Canadian activists charter a boat and sail to Amchitka Island in Alaska to attempt to stop a nuclear test; their protest becomes the first step in founding Greenpeace.

The United States bans DDT for most uses.

The Ramsar Convention on Wetlands of International Importance, Especially as Waterfowl Habitat, is signed.

An additional treaty to address the problem of oil spills, the International Convention on the Establishment of an International Fund for Compensation for Oil Pollution Damage, is signed.

The Founex Report on Development and Environment paves the way for the Stockholm conference.

The International Institute for Environment and Development, a nongovernmental organization, is founded under the name International Institute for Environmental Affairs.

1972   The Stockholm Conference on the Human Environment brings environmental concerns to the forefront of international law and policy. Representatives of 113 countries take part in the conference, which establishes environmental protection as a distinct goal of both international law and the U.N. process. The conference led to the creation of the United Nations Environment Programme (UNEP) by the General Assembly's Resolution on the Institutional and Financial Arrangements for International Environment Cooperation and to the General Assembly's adoption of the Stockholm Declaration on the Human Environment. The Stockholm Declaration, although not having any legal effect in its own right, becomes a document referred to in almost all subsequent discussions of international environmental law. Principle 21, which balances the sovereign right to develop resources within a state's territory against the duty to cause no harm outside that territory, becomes generally accepted as stating a rule of customary international law.

Several multilateral environmental treaties are adopted, including the Convention Concerning the Protection of the World Cultural and Natural Heritage, which establishes the UNESCO system of World Heritage Sites; the Convention on the Prevention of Marine Pollution by Dumping of Wastes and Other Matter; and the Convention for the Conservation of Antarctic Seals.

1973   The Convention on International Trade in Endangered Species of Wild Fauna and Flora (CITES) is opened for signature but does not enter into force until 1975; the International Convention for the Prevention of Pollution from Ships (MARPOL) is opened for signature but fails to attract support and subsequently undergoes substantial modification.

1974   The Convention for the Prevention of Marine Pollution from Land-Based Sources is opened for signature.

1975   CITES enters into force.

1976   An accident at a chemical plant in Seveso, Italy, releases toxic chemicals and contaminates a large region; public and governmental reaction to the Seveso disaster ultimately leads to the formation of a European Communities environmental directorate.

The Convention on the Prohibition of Military or Any Other Hostile Use of Environmental Modification Techniques (ENMOD) outlaws the possible use or destruction of the environment as a weapon of warfare.

Two conventions to protect the Rhine River are signed: the Convention for the Protection of the Rhine Against Chemical Pollution and the Convention Relating to the Protection of the Rhine Against Pollution by Chlorides. The former, but not the latter, is ultimately superseded by the 1999 Convention for the Protection of the Rhine.

In Bogotá, eight equatorial countries meet and declare their sovereignty over the geosynchronous orbits located 22,300 miles above their territory and reject any rules of international law to the contrary, in the Declaration of the First Meeting of the Equatorial Countries.

1977   Freeport-McMoRan begins construction on the Grasberg mine in the highlands of the Indonesian portion of New Guinea.

Two additional protocols to the Geneva Conventions of 1949 are adopted: Protocol I ("Relating to the Protection of Victims of International Armed Conflicts") and Protocol II ("Relating to the Protection of Victims of Non-International Armed Conflicts"). Protocol I contains provisions specifically addressing the problem of environmental damage during wartime; Protocol II does not, although it does include some incidental environmental protection.

Czechoslovakia and Hungary agree by treaty to construct a massive hydroelectric and navigation project along the Danube between Gabčíkovo in Czechoslovakia and Nagymaros in Hungary.

The United Nations Conference on Desertification in Nairobi, Kenya, addresses the problems of drought, loss of arable land, and expansion of deserts.

The United Nations Water Conference in Mar del Plata, Argentina, calls for the development of national water resource assessments and for national policies and plans to give priority to supplying safe drinking water and sanitation services. This leads directly to the "International Drinking Water Supply and Sanitation Decade" in the 1980s.

1978   The Convention for the Prevention of Marine Pollution from Land-Based Sources enters into force.

The International Maritime Organization (IMO) holds a Conference on Tanker Safety and Pollution Prevention. The IMO recognizes that routine operational discharge of oil from ships is a greater environmental threat than the more highly publicized oil spills. The 1973 MARPOL had attempted to address both types of pollution, as well as other vessel-source pollution but had failed to attract support. The 1978 conference adopts a protocol to MARPOL that has the effect of saving the treaty, and MARPOL, as altered by the protocol, finally enters into force in 1983.

Other international environmental agreements include the Kuwait Regional Convention for Co-operation on the Protection of the Marine Environment from Pollution.

1979   A number of multilateral environmental and natural resource agreements are concluded, including the Agreement Governing the Activities of States on the Moon and Other Celestial Bodies, the Convention on the Conservation of European Wildlife and Natural Habitats, the Convention on Long-Range Transboundary Air Pollution, and the Convention on the Prevention of Marine Pollution by Dumping of Wastes and Other Matter.

President Carter issues Executive Order 12,114, titled Environmental Effects Abroad of Major Federal Actions, imposing some environmental assessment requirements.

The World Meteorological Organization holds a World Climate Conference in Geneva. The conference concludes that the enhanced greenhouse effect from the increased buildup of carbon dioxide in the atmosphere is an urgent problem requiring international action.

The merger of several activist groups in different countries creates Greenpeace International.

1980 Canada, Norway, Sweden, and the United States ban most aerosol uses of ozone-depleting chlorofluorocarbons.

Hungarian scientists and engineers criticize the Gabčíkovo-Nagymaros project at a public meeting.

The Convention on the Conservation of Antarctic Marine Living Resources is opened for signature.

1981 The 1980s (1981–1990) are the International Drinking Water Supply and Sanitation Decade.

Hungarian journalist Janos Vargha publishes an article condemning the Gabčíkovo-Nagymaros project; the Hungarian government halts work at Nagymaros.

The Grasberg mine begins operating on the Indonesian half of New Guinea. Under the terms of an agreement between the mine's operator (Freeport-McMoRan) and the Indonesian government, Freeport-McMoRan's license to operate the mine will run until 2011.

1982 After many years of negotiation, the United Nations Convention on the Law of the Sea (UNCLOS), which includes many environmental protection provisions, is opened for signature. UNCLOS does not enter into force until 1994, although many of its provisions express, or come to be accepted as, customary international law.

The Convention on the Conservation of Antarctic Marine Living Resources enters into force.

The International Whaling Commission imposes an eighteen-year moratorium on commercial whaling, to take effect in 1985–1986. Norway, a whaling nation, objects to the moratorium in a timely fashion and is not bound by it.

The General Assembly adopts the World Charter for Nature over a single dissenting vote—that of the United States.

1983 The United States and Mexico sign an Agreement to Cooperate in the Solution of Environmental Problems in the Border Area.

MARPOL enters into force.

1984 The World Commission on Environment and Development, headed by Dr. Gro Harlem Brundtland of Norway, is formed to study and make recommendations on sustainable development.

The world's worst industrial accident occurs at a Union Carbide pesticide plant in Bhopal, India, killing nearly 4,000 people and causing lasting health problems, including blindness, for many more.

1985 The existence of a hole in the earth's protective ozone layer over Antarctica is confirmed. The ozone hole becomes a permanent feature of the atmosphere, varying seasonally in size. The world reacts quickly to the threat posed by depletion of the ozone layer; the Vienna Convention for the Protection of the Ozone Layer is signed in 1985.

French intelligence agents sink the Greenpeace ship *Rainbow Warrior,* used in protests against French nuclear tests, in a New Zealand harbor; one person is killed, and France's action is internationally condemned.

1986   An accident at a nuclear reactor at Chernobyl in the Soviet Union (now Ukraine) releases about seven tons of radioactive material into the atmosphere.

Austria agrees to loan money to Hungary for the completion of the Nagymaros portion of the Gabčíkovo-Nagymaros project, on terms beneficial to Austria.

1987   The Brundtland Report, *Our Common Future,* by the World Commission on Environment and Development, brings the term and concept of "sustainable development" to the forefront of environmental consciousness.

The Montreal Protocol on Substances that Deplete the Ozone Layer is signed.

UNEP establishes a working group to create an umbrella convention unifying existing treaties to protect biodiversity; the working group is unsuccessful, and five years later the Convention on Biological Diversity is adopted instead.

1988   Environmental activist Chico Mendes is murdered in Brazil.

The Hungarian Parliament votes to continue the Gabčíkovo-Nagymaros project.

The Toronto Atmosphere Conference calls for a 20 percent reduction in emissions of greenhouse gases by 2005. The United Nations creates an Intergovernmental Panel on Climate Change.

1989   The Basel Convention on the Control of Transboundary Movements of Hazardous Wastes and their Disposal seeks to control shipments of wastes but is seen by some developing countries as licensing the shipment of hazardous wastes from the global North to the South.

The Convention for the Prohibition of Fishing with Long Driftnets in the South Pacific is adopted.

Hungary announces free elections and a referendum on the Gabčíkovo-Nagymaros project; the Hungarian Parliament votes to abandon the project.

1990 In response to extensive pollution and human rights violations in Nigeria, Ken Saro-Wiwa founds the Movement for the Survival of the Ogoni People.

1991 Czechoslovakia begins construction of additional structures necessary to its "provisional solution," which will allow it to operate its portion of the Gabčíkovo-Nagymaros project even in the absence of the power plant to be built at Nagymaros, Hungary. The provisional solution poses a risk of environmental harm to Hungary.

Ecojuris Institute, Russia's first public interest environmental law firm, is founded in Moscow.

In the first Gulf War, Iraqi forces set fire to hundreds of oil wells in Kuwait and deliberately spill oil into the Persian Gulf. The United Nations Security Council, in Resolution 687, "reaffirms that Iraq . . . is liable under international law for any direct loss, damage, including environmental damage and the depletion of natural resources[.]" This is the first resolution of the Security Council to impose liability for environmental harm.

Canada and the United States sign an Agreement on Air Quality to resolve a long-running dispute over acid rain caused by transboundary air pollution.

Several African countries unite to address the inadequacies of the Basel Convention with the Bamako Convention on the Ban of Imports into Africa and the Control of Transboundary Movement and Management of Hazardous Wastes within Africa. Other multilateral environmental treaties of 1991 include the Espoo Convention on Environmental Impact Assessment in a Transboundary

Context and the Protocol on Environmental Protection to the Antarctic Treaty.

The Global Environment Facility (GEF) is established to provide financial assistance for programs in developing countries that will protect the global environment.

1992   Hungary informs Czechoslovakia that it is unilaterally terminating the Gabčíkovo-Nagymaros treaty. Slovakia, soon to be independent of Czechoslovakia, begins diverting water from the Danube.

In Japan, Iwasaki Corporation announces plans to build a golf course on one of the last remaining habitats of the endangered Amami Black Rabbit.

The United Nations Conference on Environment and Development is held in Rio de Janeiro, Brazil. It produces a number of documents, including the Rio Declaration, setting forth principles of sustainable development, and the detailed environmental action plan Agenda 21. The United Nations creates a Commission on Sustainable Development to follow up on the Rio conference and report on the implementation of Agenda 21.

The International Conference on Water and the Environment in Dublin adopts four guiding principles for water resource management: (1) "fresh water is a finite and vulnerable resource, essential to sustain life, development and the environment" (2) "water development and management should be based on a participatory approach, involving users, planners and policy-makers at all levels"(3) "women play a central part in the provision, management and safeguarding of water" and (4) "water has an economic value in all its competing uses and should be recognized as an economic good."

Important multilateral treaties of 1992 include the United Nations Economic Commission for Europe Convention on the Protection and Use of Transboundary Watercourses and International Lakes, the United Nations

Framework Convention on Climate Change, and the Convention on Biological Diversity.

1993 Slovakia becomes an independent nation. By special agreement, Hungary and Slovakia agree to submit the dispute over the Gabčíkovo-Nagymaros project to the International Court of Justice.

Faced with demonstrations in Nigeria's Ogoniland demanding environmental restoration and compensation, Anglo-Dutch Shell decides to cease oil operations in Ogoniland.

A federal appellate court applies the United States National Environmental Policy Act to activities of the U.S. National Science Foundation in Antarctica, but a federal trial court refuses to apply the same act to U.S. activities in Japan.

The International Court of Justice creates a Chamber for Environmental Matters; to date, the chamber has never been used.

1994 In Ogoniland, following a violent raid in which many people are killed, Ken Saro-Wiwa and eight other environmental and Ogoni rights activists are arrested.

The World Trade Organization is formed.

The United Nations Convention on the Law of the Sea enters into force.

Canada, the United States, and Mexico sign the North American Free Trade Agreement and an environmental side agreement, the North American Agreement for Environmental Cooperation. The latter agreement creates the Commission for Environmental Cooperation.

The second meeting of the Basel Convention's Conference of the Parties agrees to ban the export of hazardous wastes from Organization for Economic Cooperation and Development (OECD) countries to non-OECD countries, but the process by which the decision is reached does not comply with the Basel Convention's amendment procedure.

1995 Ken Saro-Wiwa and eight others are executed in Nigeria, provoking an international outcry from governments and environmental and human rights activists.

The third meeting of the Basel Convention's Conference of the Parties proposes that the ban on shipments of wastes from rich countries to poor ones be incorporated as a formal amendment. This ban amendment is not yet in force, and may never be.

1996 The ozone regime's ban on the production of ozone-depleting chlorofluorocarbons comes into force for industrialized countries.

The International Court of Justice issues its advisory opinion on the Legality of the Threat or Use of Nuclear Weapons, stating that the use of nuclear weapons is not necessarily forbidden by international law.

1997 The Commission on Sustainable Development reports that little progress has been made in meeting the goals of Agenda 21 in the five years since the Rio Earth Summit.

The U.N. General Assembly adopts the United Nations Convention on the Law of the Non-Navigational Uses of International Watercourses, which is then opened for signature. It has not yet entered into force.

The International Court of Justice issues its opinion in the Gabčíkovo-Nagymaros Dispute, finding rights and wrongs on both sides and stating that sustainable development is not a rule or normative principle of customary international law. In a separate opinion Vice President Weeramantry disagrees.

A federal district court in Louisiana dismisses the claim of Tom Beanal, an Amungme tribesman from the Indonesian portion of New Guinea, against Freeport-Mac-

MoRan, a U.S. mining company. Beanal had sought relief for the damage done to the Amungme homeland by the company's mining operation; he appeals the district court's decision.

The Kyoto Protocol is signed by some countries but does not enter into effect.

1998   The Rome Statute on the International Criminal Court is adopted; it enters into force in 2002. It includes some environmental crimes within its definition of war crimes.

The Rotterdam Convention for the Prior Informed Consent Procedure for Certain Hazardous Chemicals and Pesticides in International Trade is signed; it enters into force in 2004.

1999   During the war between the North Atlantic Treaty Organization (NATO) and the Federal Republic of Yugoslavia (now Serbia-Montenegro), NATO forces repeatedly bomb a chemical storage and manufacturing complex at Pančevo. Yugoslavia claims that the bombing of the Pančevo complex is an environmental war crime. A UNEP task force concludes that the damage, while significant, is not catastrophic.

Yugoslavia files lawsuits before the International Court of Justice against Belgium, Canada, France, Germany, Italy, the Netherlands, Portugal, Spain, the United Kingdom, and the United States, alleging, among other things, environmental war crimes. Two of the suits (against Spain and the United States) are promptly dismissed for lack of jurisdiction, but the other eight remain pending.

The United States Court of Appeals for the Fifth Circuit upholds the district court's dismissal of Tom Beanal's complaint against Freeport-McMoRan.

In Seattle, the anti-globalization movement comes of age with the Battle of Seattle, successfully disrupting World Trade Organization talks. Some protesters are motivated by environmental concerns, sparking a debate about the relationship between trade, wealth, and the environment.

The Convention for the Protection of the Rhine is concluded, repealing the 1963 Agreement Concerning the International Commission for the Protection of the Rhine against Pollution and its Additional Agreement of 1976, as well as the 1976 Convention for the Protection of the Rhine against Chemical Pollution. The 1976 Rhine Chloride Convention remains in force, however.

2000   A cyanide spill from a mine at Baia Mare, Romania, jointly operated by an Australian company and the Romanian government, causes severe environmental damage to the Tisza River, especially in Hungary.

The Office of the Prosecutor of the International Criminal Tribunal for the Former Yugoslavia concludes that NATO's bombing of the Pančevo complex was not an environmental war crime and declines to bring charges.

The Cartagena Protocol on Biosafety to the Convention on Biological Diversity addresses genetically modified organisms; it enters into force in 2003.

The Bonn Guidelines on Access to Genetic Resources and Fair and Equitable Sharing of the Benefits Arising out of their Utilization attempt to address problems of bio-piracy and the rights of indigenous peoples in their traditional knowledge.

2001   The United States announces that it will not implement the Kyoto Protocol and withdraws from the Kyoto Protocol process.

The Stockholm Convention on Persistent Organic Pollutants is opened for signature; it entered into force in 2004; it addresses chemical pollutants that for the most part are small in total quantity but highly toxic, including pesticides, dioxins, and polychlorinated biphenyls (PCBs).

The Permanent Court of Arbitration adopts rules for the arbitration of environmental disputes.

The World Bank releases its environmental manifesto "Making Sustainable Commitments: An Environmental Strategy for the World Bank."

2002   The World Summit on Sustainable Development in Johannesburg, South Africa, adopts the Johannesburg Plan of Implementation.

The Rome Statute on the International Criminal Court is adopted; it enters into force in 2002.

2003   The Cartagena Protocol on Biosafety to the Convention on Biological Diversity enters into force.

2004   The Rotterdam Convention for the Prior Informed Consent Procedure for Certain Hazardous Chemicals and Pesticides in International Trade enters into force, as does the Stockholm Convention on Persistent Organic Pollutants.

Russia ratifies the Kyoto Protocol.

2050   Year by which concentration of ozone-destroying chlorofluorocarbons in the ozone layer is expected to decline to pre-1980 levels, according to a United Nations Industrial Development Organization estimate.

5

Biographical Sketches

Ultimately, the force that drives international environmental law is people: People who want to protect the environment, people who want to develop natural resources, people caught in the middle. Nations may make international law, but they too are composed of individuals—and they act in response, and sometimes in opposition, to the desires of those individuals.

The individuals described in this chapter have all played some role in influencing international environmental law and policy. Their life stories are as varied as those of any people on earth: One was for many years the prime minister of one of the wealthiest nations on earth, another the illiterate daughter of a rubber-tapper in the forests of Brazil's Amazon region. Akiko Domoto and Janos Vargha both began as journalists, one on television in a wealthy democratic country, the other in print in an "Iron Curtain" country under one-party rule. Some are environmental lawyers, like Vera Mischenko, or became lawyers after careers in grassroots activism and electoral politics, like Harrison Ngau Laing. Others, like Ken Saro-Wiwa and Bjørn Lomborg, have been university teachers and authors, of very different sorts. Ken Saro-Wiwa, a successful businessman and internationally famous author, risked—and lost—his life in an attempt to protect the environment and people of the Niger Delta. Chico Mendes, the colleague and mentor of Marina Silva, also became a martyr to the cause of environmental protection. And the world has grown smaller: Tom Beanal, a tribal elder from the highlands of New Guinea, has brought his environmental struggle to a federal courtroom in New Orleans, Louisiana.

157

International environmental law today has changed and evolved because of these people, and countless others like them. The Niger Delta may still be polluted, and the mining Tom Beanal opposed may still be going on, but international corporations step far more carefully than before. The commission chaired by Norwegian Prime Minister Gro Harlem Brundtland set forth the idea of sustainable development as it is understood today and led to the Rio Summit at which that understanding was formalized and further developed.

Domestic law has changed as well. Large areas of Brazilian forest have been declared protected areas because of the efforts of Ms. Silva. The environmental laws of Russia, Japan, and Malaysia have changed and are still changing because of Ms. Mischenko, Ms. Domoto, and Mr. Ngau—and because of the international nature of the environmental movement and the ongoing exchange of ideas among environmentalists worldwide.

Tom Beanal (1947–)

Tom Beanal is a tribal elder of the Amungme people of West Papua (formerly known as Irian Jaya, before that as Irian Barat, and before that as West Irian), a province of Indonesia. It shares the island of New Guinea with the independent nation of Papua New Guinea. New Guinea contains one of the highest concentrations of human diversity—linguistic, cultural, and ethnic—in the world; the majority of the indigenous inhabitants of West Papua do not share a language, ethnicity, religion, or culture with the majority of non-Papuan Indonesians.

Tom Beanal was born in the highlands of New Guinea in 1947; at the time West Papua was under Dutch colonial rule. When Indonesia became an independent nation two years later, West Papua did not join it; it remained under Dutch rule until 1962. During the last ten years of that time, the Netherlands made plans to transfer power to an independent West Papuan government; independence was scheduled for 1970. In 1961, however, Indonesia began an armed conflict against the Dutch administration; in late 1962 the parties agreed to hand West Papua over to a temporary U.N. administration, which would then hand the territory over to Indonesia until a plebiscite could be conducted to determine the will of the West Papuan people.

Indonesia took control of the territory in May 1963, and conducted the "plebiscite" in 1969. Only a few hundred people, selected by the Indonesian military, actually voted; the two international observers left after only 200 or so votes had been cast. The international community was nonetheless satisfied that this procedure satisfied the requirements of an "Act of Free Will" as established by the United Nations, and West Papua became a part of Indonesia (Human Rights Watch, 2000).

West Papuan groups have engaged in both armed and nonviolent resistance to Indonesian rule more or less continuously since Indonesian rule began; partly as a result, the Indonesian military has always maintained a large presence in the territory. The Indonesian government has also encouraged the migration of a large number of other Indonesians into West Papua, so that the indigenous Papuans are now a minority.

In 1967, the Indonesian government signed an agreement giving a U.S. company, Freeport-McMoRan, exclusive mining rights to an area of West Papua larger than many U.S. states. The agreement was reached, rather brazenly, before the plebiscite had been conducted and gives Freeport-McMoRan a thirty-year mining license from the date of the opening of the mine. Construction on the mine began in 1977, and the thirty years began to run when the mine opened in 1981.

The area of the Freeport-McMoRan mining concession was occupied by the Amungme people. The mountain where mining began was sacred to the Amungme. Tom Beanal explained years later in a commencement address at Loyola University in New Orleans that:

When the earth was first created, it is believed that the Amungme people occupied land which was still swamp. The story goes that there was a mother with four children, two boys and two girls. They lived in the middle of the swamp, where there was a dry land.

One day the dry season came. There was famine and many people died. This also affected the mother and her four children. They began to suffer from hunger, when the food they had stored was used up.

The mother said to her children, "Instead of all of us dying, it is better if just I die." She ordered her children to kill her. She asked them to cut off her head and throw it to the north. She asked for her body to be cut into two, with the right side being thrown to the east and the left side to the west. Her feet were to be thrown towards the river so that they would be brought south by the current. Her children carried out this task with heavy hearts.

After they had done what their mother had asked, the four children fell asleep. When they awoke, they were surprised to see a mountain in the north, where they had thrown their mother's head. In the east and west there grew a great garden with all kinds of things to eat. In the south as well, there was a broad expanse of land.

From the point of view of the Amungme, the mountain Freeport-McMoRan was mining was the head of their ancestress, who had sacrificed herself so that her descendants might live. The mountain no longer exists; it has been entirely excavated, and where it once stood is now the world's largest operating open-pit mine.

Sacrilege was not the only cause of conflict between the Amungme and Freeport-McMoRan. Amungme villages were displaced. Game animals vanished, and water became polluted. Sago palms, a staple food source, have become scarce. Indonesian soldiers and mining company employees are alleged to have raped, tortured, and murdered Amungme people.

Tom Beanal became one of the most outspoken critics of Freeport-McMoRan; he succeeded in attracting the attention of the media and environmental and human rights activists. In 1996 he sued Freeport-McMoRan on its home ground, in the U.S. District Court for the Eastern District of Louisiana. Some confusion surrounded the initiation of the litigation; reports in the media indicated that Beanal had not authorized the filing of the suit and included copies of a letter purportedly from Beanal instructing his attorney to dismiss it. The authenticity of the letter, though, was called into question: the letter was in English, a language in which Tom Beanal could not write, rather than in Bahasa or Amungme.

Tom Beanal participated in the litigation, however; he gave a deposition in Djakarta and testified before the court in New Orleans. The litigation was ultimately unsuccessful in achieving its ostensible objective; the court eventually found that Beanal had failed to state a cause of action for a violation of international environmental law, and the finding was upheld on appeal. While the result was unwelcome to human rights and environmental advocates in the United States, the litigation may have achieved some success in a broader sense. Freeport-McMoRan, which previously had operated in comfortable obscurity, became a focal point for protesters worldwide. And Indonesia was in a time of crisis: The independence of East Timor, along with the activities of separatist movements, outbreaks of ethnic violence across Indonesia, and Suharto's fall and the subsequent instability accompanying Indonesia's transition to democracy, led to widespread speculation that Indonesia might be about to go the way of the Soviet Union and Yugoslavia, breaking up along ethnic lines (Kingsbury and Aveling, 2003). If that were to happen, Freeport-McMoRan would be in a vulnerable position. Mines cannot be moved like other businesses; they must either operate where they are or shut down. Freeport-McMoRan would have to operate under an independent West Papuan government—one that could be expected to be unfriendly to Freeport-McMoRan, and one in which Tom Beanal himself, as one of the island's best-known indigenous leaders, might occupy a high position.

Presumably to protect itself against both international criticism and the possibility of West Papuan independence, Freeport appointed Beanal to its board of directors. The move led some West Papuans to wonder whether Beanal had sold out. Beanal delayed for two years before accepting appointment to Freeport-McMoRan's board; finally, he said, he realized "this post is not a reward, it is my duty since it involves the use of our land . . . . I should have assumed this post in 1967."

Beanal's appointment was unpopular with the Indonesian military; relations between Freeport and the military seem to have soured somewhat, and the military, rather than Papuan bandits, has been blamed by some for an attack that killed two Americans (Murphy, 2002). Meanwhile Tom Beanal has become an icon. His description of the plight of Amungme is equally applicable to indigenous peoples everywhere: "We want to develop ourselves, not just be developed."

Gro Harlem Brundtland (1939–)

The idea of sustainable development, one of the guiding principles of modern international environmental law, underwent considerable evolution in the years leading up to the Earth Summit at Rio de Janeiro. Much of that evolution took place as a result of the work of the World Commission on Environment and Development, better known as the Brundtland Commission after its chair, Dr. Gro Harlem Brundtland.

Dr. Brundtland spent the earliest years of her childhood in Norway under Nazi occupation; Germany invaded Norway when she was just over a year old. Shortly after the war, when she was ten years old, her family moved to the United States. Her education covered both countries, with degrees in medicine from the University of Oslo in 1963 and in public health from Harvard in 1965. Since returning to Oslo in 1965 she has spent her entire career in public service. As an official with the Norwegian Ministry of Health, she became increasingly aware of and concerned about the link between health and the environment; in 1974 she became Norway's minister for environmental affairs, and in 1981, its prime minister—the youngest person and first woman ever to hold that post in Norway (Brundtland, 2002).

Many people outside Norway might be hard pressed to name even one Norwegian prime minister, other than the notorious traitor Vidkun Quisling. It is her international work that has made Dr. Brundtland's name familiar to environmentalists everywhere. In 1983, the General Assembly, in a resolution titled "Process of Preparation of the Environmental Perspective to the Year 2000 and Beyond," urged the establishment of a commission:

(a) To propose long-term environmental strategies for achieving sustainable development to the year 2000 and beyond;

(b) To recommend ways in which concern for the environment may be translated into greater co-operation among developing countries and between countries at different stages of economic and social development;

(c) To consider ways and means by which the international community can deal more effectively with environmental concerns;

(d) To help to define shared perceptions of long-term environmental issues and of the appropriate efforts needed to deal successfully with the problems of protecting and enhancing the environment, a long-term agenda for action during the coming decades, and aspirational goals for the world community.

The Brundtland Commission worked on these issues throughout the mid-1980s, paving the way for the 1992 Rio Summit—the United Nations Conference on Environment and Development. The commission published its report, under the title *Our Common Future,* in 1987, and officially reported it to the United Nations General Assembly in 1989, setting in motion the organization of the Rio Summit. The hard-copy version of *Our Common Future* is available from Oxford University Press; the report is also available as a free download. It sets out the principles of sustainable development as generally understood today. The report's capsule definition of sustainable development—"Sustainable development is development that meets the needs of the present without compromising the ability of future generations to meet their own needs"—is the ancestor of Rio Declaration Principle 3's statement of intergenerational equity as an element of sustainability: "The right to development must be fulfilled so as to equitably meet developmental and environmental needs of present and future generations."

Dr. Brundtland's efforts on behalf of the environment and public health did not end with the Brundtland Report, however. In addition to serving as head of Norway's Labor Party until 1992 and as prime minister, intermittently, until 1996, she became director-general of the World Health Organization (WHO) in 1998 and served in that post until 2003. Of the WHO, she says "I see WHO's role as being the moral voice and the technical leader in improving health of the people of the world, ready and able to give advice on the key issues that can unleash development and alleviate suffering. I see our purpose to be combating disease and ill health, promoting sustainable and equitable health systems in all countries" (World Health Organization, 2004). At WHO, she won acclaim for coordinating the successful international effort to contain Severe Acute Respiratory Syndrome (SARS).

Akiko Domoto (1932–)

Japan is among the world's wealthiest and most developed countries; with a high population density and a high level of industrialization, the space left for nature has grown smaller and smaller. The struggles of environmentalists to preserve the quality of Japan's human and natural environment would be familiar to any developed-country environmental activist: passive resistance to prevent the construction of a golf course on the last remaining habitat of the endangered Amami Rabbit, lawsuits and letter-writing campaigns to demand national-government investigation of water pollution levels, lobbying local governments to prevent draining and filling of wetlands. For decades, television personality and director Akiko Domoto was a prominent voice in Japan's environmental and feminist movements; in 1989 she was a natural choice for an opposition party looking for a parliamentary candidate (*Business Week Online*, 2001). As a member of the House of Councilors (the upper house of Japan's Diet, or parliament), she worked tirelessly on feminist, environmental, and human rights issues; she was instrumental in the passage of legislation banning domestic violence, gender discrimination, and child prostitution. She also became involved in the struggle to save the Amami Black Rabbit, an endangered species that has acquired iconic status among both the environmental and the anti-environmental movements in Japan, much like the spotted owl in the United States. Ms. Domoto explained:

> The Amami Rabbit is a short-eared, black primitive species of rabbit which is found nowhere else in the world but two islands in Japan, the Amami Islands in Kagoshima Prefecture. Its closest relatives live in Mexico and South Africa. Listed in the IUCN Red Data Book as endangered, it is also (supposed to be) protected as a Japanese "Special Natural Monument."
>
> In 1977 a survey revealed a population of 6,000 rabbits. Today, there is estimated to be as few as 1,000 Amami Rabbits remaining in only three areas of vital, high-density habitat. The primary reason for this dramatic population decline is habitat destruction, mostly due to the development of the Amami tourism industry.
>
> In 1992 the Iwasaki Company announced plans to build a golf course on the richest of these three remaining habitats. The proposed golf course will be a 170 hectare, 18-hole course located on a promontory overlooking the Pacific Ocean.

Because the rabbits were nocturnal, she explained, they would be likely to be sleeping in their burrows during the day and unable to escape the bulldozers. The local government's provision for protection of the rabbits was to begin construction at the ocean's shore and move inland, "thus allowing the rabbits to flee from the path of construction into the surrounding forest."

Environmental activists had opposed the golf course with letter-writing campaigns and demonstrations. Attempts by the national government to intervene were fruitless; although the government attempted to conduct an environmental impact assessment, it was allegedly barred from doing so by the landowner, Iwasaki Corporation, which then hired a private consultant to conduct an environmental survey. The local government, Ms. Domoto stated, concluded "that according to its review of Iwasaki's survey results, the impact of the golf course on the Amami Rabbit would be limited." Even a plea from the Duke of Edinburgh proved fruitless. In addition to their work within the Diet, Ms. Domoto and several other legislators formed a "Parliamentarian's League for the Conservation of the Amami Black Rabbit," attracting considerable media attention.

In the meantime, and in part as a result of the publicity, a revolutionary environmental lawsuit, in terms of existing Japanese environmental law, was brought: the Amami Rights of Nature lawsuit. Volunteer lawyers filed a complaint naming the rabbit and four other species, along with some humans, as plaintiffs. The lawsuit, which was ultimately dismissed, invoked both foreign, especially U.S., environmental law and the traditional Japanese concept that humans and nature coexist as part of a single society, and that humans are not the only ones with rights in that society (Kagohashi, 2002).

Although the golf course developers won all the battles, they lost the war: The project was eventually abandoned as a result of the combination of litigation, national government opposition, and unfavorable media coverage and popular outrage both within Japan and internationally.

Ms. Domoto is no longer a member of the Diet; in 2001 she was elected governor of Chiba Prefecture, near Tokyo. Chiba has suffered from toxic-waste dumping, and its coastal wetlands have been severely damaged by land-reclamation projects. Ms. Domoto has take on the formidable task of cleaning up Chiba, stopping the dumping, and saving its last remaining tidal wetlands, while at the same time ensuring its prosperity.

Bjørn Lomborg (1965–)

Just as the international environmental activist movement has made saints of some, it has made demons of others; at the moment, one of the chief demons is a once-obscure professor of statistics from Denmark named Bjørn Lomborg. In 1998 Lomborg published four articles in *Politiken,* a Danish periodical, questioning the generally accepted belief that the quality of the human environment is deteriorating. Although the articles resulted in a nationwide debate within Denmark, few people outside Scandinavia read Danish; Lomborg's articles might have had little effect on the outside world had the ideas in them not been expanded upon and published in book form, first in Danish as *Verdens Sande Tilstand (The True State of the World)* and later in English as *The Skeptical Environmentalist: Measuring the Real State of the World.* (The title is a play on the annual *State of the World* report published by the Worldwatch Institute.) The 2001 publication of *The Skeptical Environmentalist* made Lomborg an overnight celebrity, adored by anti-environmentalists and abhorred by Greens everywhere. The book was denounced by groups as various as the Danish Committee for Scientific Dishonesty and the Union of Concerned Scientists, as well as publications such as *Scientific American* and *Nature.* An Internet search for "Skeptical Environmentalist" or "Bjørn Lomborg" turns up thousands of pages of argument pro and con, often in the form of barely coherent rants. The Union of Concerned Scientists wrote that, "Lomborg's book is seriously flawed and fails to meet basic standards of credible scientific analysis" (Union of Concerned Scientists, 2003). Some of Lomborg's detractors can barely contain themselves; a review in the usually staid British journal *Nature* stated that "the text employs the strategy of those who, for example, argue that gay men aren't dying of AIDS, that Jews weren't singled out by the Nazis for extermination, and so on" (Pimm and Harvey, 2001). Websites attack his qualifications to teach statistics, his doctoral dissertation, and other aspects of his personal life. Environmentalist author Mark Lynas threw a pie in Lomborg's face at a Borders bookstore in Oxford. "I wanted to put a Baked Alaska in his smug face," said Lynas, "in solidarity with the native Indian and Eskimo people in Alaska who are reporting rising temperatures, shrinking sea ice and worsening effects on animal and bird life" (Cardwell, 2004). Danish biologist Kåre Fog has dedicated a website, http://www.lomborg-errors.dk/, to pointing out errors in the book: The website also, in its own words, "gives information on cases and activities related to Bjørn Lomborg, attempts to describe his methods, and attempts to document his dishonesty." Many of his most enthusiastic admirers, on the other hand, do his cause and his reputation more harm than good when, in the same paragraph in which they declare their devotion to his work, they also declare their undying faith in cold fusion, space colonization, and the Libertarian Party.

This is an amazing amount of attention to devote to the author of a single book that only a small number of people have actually read. The furor generated by Lomborg's book is all the more surprising because the basic thesis—that the human environment is not only not deteriorating, but in many ways improving—is not new. Lomborg claims as his own inspiration the similar work of the late economist Julian Simon; in the introduction to his book, Lomborg says that he started out investigating Simon's work with his students, expecting that he and his students would "show that most of Simon's talk was simple, American rightwing propaganda." And the title of an earlier work by Simon, *The State of Humanity*, is also a play on Worldwatch's *The State of the World* and a possible additional inspiration for Lomborg's title.

But Simon wrote primarily for the converted. The dozen books similar to Lomborg's that are for sale at any given moment are also designed, like so much political discourse these days, to strengthen the convictions of the already-convinced. Much of what is written on the environmentalist side suffers from the same shrillness. Environmentalists and anti-environmentalists shrieking at each other may relieve some tensions, but it does little to extend actual awareness of environmental problems and solutions.

Lomborg's book has attracted more attention than many similar works precisely because it is less intolerantly anti-Green. Lomborg's basic thesis is that things are getting better, not worse; underlying and closely related to this is the idea that resources devoted to environmental protection are being misallocated. Opposing and banning genetically modified food, for example, requires time, energy, and money that might be spent improving prenatal nutrition, saving far more lives. Lomborg's ideas make sense not only to anti-environmentalists and the undecided, but to many (although, obviously, by no means all) environmentalists. Instead of tapping into the anti-environmentalist resentment that has been previously exploited by other authors, Lomborg challenged environmentalists to examine their preconception. His work may well be flawed as seriously as his detractors claim, but it has had a salutary effect on the environmentalist movement as a whole: It has forced a careful, critical reexamination of the facts and figures upon which so much of our environmental policy is based.

Since the English publication of his book by Cambridge University Press, Lomborg himself has not been teaching; for the past three years he has been on leave from the University of Aarhus. From February 2002 until July 2004, he acted as director of Denmark's national Environmental Assessment Institute. In December 2003, he achieved a vindication of sorts: The Danish Ministry of Science, Technology, and Innovation invalidated the earlier finding of the Danish Committee on Scientific Dishonesty that Lomborg's book was "objectively dishonest" and "clearly contrary to the standards of good scientific practice." In March 2004, the Committee on Scientific Dishonesty announced that it would not reopen the case (Lomborg, 2004). Lomborg has published a great deal of material online, much of it in response to criticisms of *The Skeptical Environmentalist,* and in 2004 published a follow-up book, *Global Crises, Global Solutions.*

Vera Mischenko (1953–)

Seven decades of Communist Party rule left the countries of the former Soviet Union with a legacy of environmental disaster. An ideology that viewed nature as an opponent and gigantic industrial projects as desirable in their own right regardless of their utility were among the causes. State-controlled media suppressed news of pollution and environmental disasters. And government suspicion of political activism prevented the formation of independent environmental groups (French, 1990).

The fall of Communism and the breakup of the Soviet Union made matters better in some ways, but worse in others. Hundreds of environmental groups formed; the media enthusiastically reported on the environmental sins of the old regime. But corruption, financial problems, and lack of qualified personnel have left the government unable to enforce its environmental laws across a country nearly twice the size of the United States, while at the same time increased trade with the outside world has encouraged poorly policed exploitation of timber, oil, and fisheries resources, among others.

In 1991, confronted with a government both unable and unwilling to tackle the task of protecting Russia's environment on its own, Vera Mischenko and others founded the Ecojuris Institute, the country's first public interest environmental law firm, in Moscow. Ms. Mischenko is an attorney with a doctorate in Environmental and Natural Resources Law from the prestigious

Moscow State University, alma mater of Mikhail Gorbachev. The work of the Ecojuris Institute superficially resembles that of an environmental law firm in the United States: It bring suits to protect old-growth forests and the indigenous peoples who live in them, to stop or force alterations to destructive construction projects, to demand safety precautions for oil-drilling operations, and to protect the populace from health problems caused by water and air pollution, lobbying for changes in environmental laws. But beyond these superficial resemblances lie the difficulties inherent in trying to enforce laws in a country with a barely working legal infrastructure and with environmental problems so severe that problems that might be considered national catastrophes in other countries would be viewed as mere vanity complaints in Russia (Feshbach and Friendly, 1992).

Sakhalin Island lies a quarter of the way around the globe from Moscow, north of Japan in the Sea of Okhotsk, roughly as close to Seattle or to New Guinea as to Moscow. The size of South Carolina, it holds just 700,000 people in comparison to the Palmetto State's 4,100,000: an ethnic Russian majority, a substantial ethnic Korean minority, and smaller numbers of various tribal peoples, including Ainu, Nivkhs, Orochons, and Evenkyi. The indigenous peoples suffered greatly from the use of Sakhalin Island as a penal colony in the nineteenth century and today number under 10,000. Sakhalin Island's fisheries provide Russia with most of its domestically consumed seafood. Its small population, remote location, and isolation have left much of its natural environment untouched; its waters and forests are home to rare mammals and birds and are seasonally visited by migrating marine mammals.

Although Sakhalin does attract some Russian and Japanese tourists, its potential for tourism, ecologically sound or otherwise, is limited by the difficulty of getting there. What the island does have is oil. Since the breakup of the Soviet Union, Sakhalin has attracted oil companies from around the world. After the Russian government issued a decree waiving the environmental impact assessment requirement for a proposed Exxon drilling operation, Ms. Mischenko filed a suit to block the decree and require the assessment. Ultimately, Russia's highest court agreed that the environmental impact assessment should be required (Goldman Environmental Prize, 2000). While this might not seem an unusual result in many countries, it represented a dramatic step forward in the enforcement of environmental law in Russia.

The islanders themselves had mixed reactions. Sakhalin Island is not wealthy, and the oil industry may bring jobs and royalties. On the other hand, the people of Russia are among the most environmentally conscious in the world, perhaps as a result of the eco-disasters of the communist era. Then there are the fisheries. Natalia Barannikova of Sakhalin Environment Watch, a local environmental group, points out that "All this coastline is an area of commercial fishing. Up to 1,000 vessels can be simultaneously fishing in the Tartar Straits in summer" (Santana, 2003). A single oil spill could deprive all of those vessels of their catch and tens or hundreds of thousands of islanders of their livelihood. On the island itself, oil pipelines would cross streams and lakes that also support valuable commercial, sport, and subsistence fisheries, presenting further opportunities for harm to those fisheries.

The problems of Sakhalin Island are the problems of the entire country in miniature; the islanders and the Russian government must figure out how best to balance development needs against environmental needs, how to balance some types of development against others, and how to avoid the mistakes of the past. Ms. Mischenko and Ecojuris will not lack for work in the years to come.

Harrison Ngau Laing

The indigenous peoples of the Malaysian state of Sarawak on the island of Borneo, collectively referred to as Dayaks, include among their number the Penan tribe, some of Asia's last hunter-gatherers. The Penan inhabit tropical rain forests that are also the source of valuable hardwood.

While Malaysia as a whole stands somewhere on the fuzzy border that defines developing nations from developed ones, there are great disparities in levels of development within the country. The modern cities of peninsular Malaysia, with two of the world's five tallest buildings, belong to the developed world. Sarawak, with an area roughly equal to that of all of peninsular Malaysia but with less than one-eighth the population, belongs to the developing world. The population mix is quite different from that of peninsular Malaysia; Malays and ethnic Chinese make up 21 percent and 29 percent of Sarawak's population, respectively, and there is only a tiny ethnic Indian minority. The remaining half of the population belong to Sarawak's indigenous tribal groups, of which the most numerous are the Ibans, the Melanau, and the Bidayuhs. Many other tribal groups, including the few hundred Penan, are collectively referred to as "Orang Ulu," or "upriver people." Other Orang Ulu groups include the Kayan, Kelabit, and Kenyah tribes, but only the Penan maintain a hunter-gatherer lifestyle. The million or so indigenous people of Sarawak are divided into speakers of forty-five different languages; English is widely used, as is Bahasa Sarawak, a Creole variant of the national Bahasa Malaysia language (*Sarawak Online*, 2004).

Malaysia is a major timber exporter—for two decades, it has led the world in timber exports, despite a ban on the export of raw logs from peninsular Malaysia and, more recently, from the state of Sabah, adjoining Sarawak on the island of Borneo. The largest part of Malaysia's timber exports comes from Sarawak; during the 1980s Sarawak provided 40 percent of Malaysia's log exports. At that time logging companies were clearing five square miles of Sarawak's forest each day; no other part of the world was being deforested as quickly.

Harrison Ngau Laing was not a Penan but a Kayan. He was appalled both by the deforestation of Sarawak and the resulting plight of the Penan people, who depended on the forests. He led the Penan people and members of the Malaysian Friends of the Earth (Sahabat Alam Malaysia) in blockades of logging roads. In a state in which the minister of the environment was also the owner of a large lumber company, this was even riskier than similar actions in California or British Columbia. Beginning in October 1987, Ngau was placed under house arrest for nearly two years; he also spent two months in jail under Malaysia's Internal Security Act. He was forbidden to make statements to the press, attend political meetings, or hold a post in any organization (Goldman Environmental Prize, 1990).

Ngau's efforts and the plight of the Penan attracted international attention. In 1988 Sahabat Alam Malaysia, its founder S. Mohammed Idris, the Penan people, and Ngau were awarded the Right Livelihood Award, or Alternative Nobel Prize; in 1990 Ngau was awarded the Goldman Environmental Prize. The additional media coverage brought by the award was sufficient to protect Ngau himself from persecution, to some extent, if not to halt the logging. Ngau used the prize money to run for, and win, a seat in Malaysia's Parliament. He was instrumental in the successful struggle to stop the construction of an environmentally destructive hydroelectric project in Sarawak. Ngau is no longer a member of Parliament; recently he has represented indigenous people in suits to protect their land and environment against logging companies, agribusiness interests, and others. He is also the program coordinator for Lasan Patapan Adat Dayak Sarawak, a human rights and legal aid project for the indigenous peoples of Sarawak. Logging in Sarawak continues, although many Penan lands have been set aside as biosphere reserves.

Ken Saro-Wiwa (1941–1995)

In the Niger Delta of Nigeria lies Ogoniland, home to the Ogoni people. Ogoniland was once a pleasant, productive land, dotted with farms and fisheries. Into this world, Ken Saro-Wiwa was born in 1941, the son of Widy and Jim Beeson Wiwa. His father was a successful entrepreneur and local chief; Saro-Wiwa was an excellent student and child prodigy, entering the Government College in Umuahia at the age of thirteen. Umuahia had produced the celebrated Nigerian writers Chinua Achebe and Elechi Amadi; but, although Saro-Wiwa edited and wrote fiction for a student magazine, it was not immediately obvious that he would follow in their footsteps. Instead, he went on to graduate from the University of Ibadan and become a college teacher. During Nigeria's disastrous civil war in the late 1960s, he chose the government side and became an administrative official, but in 1973, he was fired for favoring autonomy for the Ogoni people. He then went into business, as his father had, and was quite successful (Greenberg, 1990).

The cause of Nigeria's civil war was the same factor that provided the root of Nigeria's (relative) wealth, the basis of its legendary corruption, and the ultimate deterioration of the Ogoni way of life: oil. Royal Dutch Shell had found oil in Ogoniland in 1958, while Nigeria was still under British rule. The oil industry is environmentally destructive even in countries with well-designed and strictly enforced environmental laws; British Nigeria in 1958 was not such a country, nor did the situation improve after independence in 1960. By the 1970s, the fisheries of Ogoniland were vanishing, and the farmland was becoming unusable. The Ogoni themselves received neither revenue nor jobs from the oil beneath their land; only 2 percent of Shell's Nigerian workforce was Ogoni (Watts, 1997).

The deterioration was gradual, however, and decades passed before a crisis was reached. In the meantime, Saro-Wiwa became one of Nigeria's best-known writers. In 1972, he had published a radio play, *The Transistor Radio*. Because few people read Ogoni and English is Nigeria's national language, he wrote in English; this made his work instantly accessible, not only to Nigerians, but to much of the world. His output was varied; for example, he wrote and produced over 150 episodes of a children's television show, *Basi & Company*, before the show was cancelled in 1992. In 1985, he published his first novel, *Sozaboy: A Novel in Rotten English*, and a book of poetry, *Songs in a Time of War*. Both were based on his experience in the civil war; they brought him not only national but international attention.

During the 1980s, the environment of Ogoniland and the plight of the Ogoni people continued to worsen. In 1990, Saro-Wiwo founded the Movement for the Survival of the Ogoni People (MOSOP); a radical group affiliated with MOSOP allegedly engaged in an ecotage campaign against Shell's operations in Ogoniland. At the time, although the Ogoni people were suffering, Saro-Wiwo was neither poor nor personally oppressed; he was a wealthy businessman who had sent five children to school in England (where one had died, tragically, playing rugby at Eton) and an internationally famous writer; unlike many other indigenous activists, his fame was not the result of his activism. Fame was an asset, however, and he used it in his struggle to protect the Ogoni people and their environment. In 1991 and 1992, he published *Nigeria, The Brink of Disaster* and *Genocide in Nigeria*, attacking the corrupt Nigerian government, Shell, and British Petroleum for their activities, especially in Ogoniland. It was in 1992 that the government cancelled the children's television show *Basi & Company*. More than $30 billion worth of oil had been extracted from Ogoniland since 1958; in January 1993, Saro-Wiwa led more than a quarter of a million Ogoni in a march to demand a share in future oil revenues, restoration of the delta environment, and compensation for damage already done. Instead, in 1993 Shell decided to cease its operations in Ogoniland altogether. Throughout this period Saro-Wiwa was detained or imprisoned several times; in 1994, however, the government apparently decided to crush the Ogoni autonomy movement. The Nigerian army invaded Ogoniland, widespread human rights violations by the army were reported, including rape, looting, and execution of civilians without trial. Saro-Wiwa and eight other alleged MOSOP leaders were arrested during a May 1994 raid on Saro-Wiwa's home; they were subsequently charged with the murder of four Ogoni chiefs.

During the raid, several of Saro-Wiwa's family members and many others were killed.

Although the facts at this point are probably impossible to determine, Saro-Wiwa, an outspoken advocate of nonviolent resistance, was almost certainly innocent. Governments around the world denounced his arrest, and Amnesty International declared him a prisoner of conscience. In October 1995, Saro-Wiwa and his codefendants were convicted of murder by a military tribunal. World leaders from divergent political and cultural perspectives united in denouncing the trial as a sham and urging that the defendants' lives be spared. The defendants' lives were not spared; they were hanged in Port Harcourt on November 10, 1995. South African President Nelson Mandela recommended that Nigeria be expelled from the British Commonwealth; British Prime Minister John Major called the trial "judicial murder"; the U.S. ambassador to the United Nations, Madeleine Albright, joined with others in condemning the executions in discussion before the Security Council. Nigeria was in fact suspended from the Commonwealth, and the U.N. General Assembly passed, by a vote of 101 in favor to 14 against, with 47 abstentions, a resolution condemning Nigeria's action.

From his prison cell, Saro-Wiwa wrote of the plight of the Ogoni and of the Nigerian populace that, "Ultimately the fault lies at the door of the British government. It is the British government which supplies arms and credit to the military dictators of Nigeria, knowing full well that all such arms will only be used against innocent, unarmed citizens." A more immediate target for much of the world's outrage was Shell. Counsel for Shell observed the trial, and Shell's position could most charitably be described as ambivalent. Two days before the executions, the chairman of Royal Dutch Shell had written a letter requesting clemency; on the same day, the managing director of Shell Petroleum Development in Nigeria reportedly wrote a letter stating, "We believe that to interfere in the processes either political or legal here in Nigeria would be wrong" (*Third World Traveler*, 2004). In his final statement to the military tribunal Saro-Wiwa declared that:

> Shell is here on trial and it is as well that it is represented by counsel said to be holding a watching brief. The company has, indeed, ducked this particular trial, but its day will surely come and the lessons learnt here may

prove useful to it for there is no doubt in my mind that the ecological war that the company has waged in the Delta will be called to question sooner than later and the crimes of that war be duly punished. The crime of the company's dirty wars against the Ogoni people will also be punished (Greenpeace, 1996).

While in prison Saro-Wiwa received the Goldman Environmental Prize; in a message smuggled to Richard and Rhoda Goldman from his cell, he wrote:

Only recently, twenty-seven years and more after my initial protest, did Shell finally agree that it has polluted the Ogoni environment, among others. The company has now launched what it calls a "major environmental survey" of the Niger Delta, of which the Ogoni strip is part. In launching the survey, Shell condemned "recent politicized and emotive campaigning" in defense of the region.

I submit that we have every reason to be emotional in our struggle for the sanctity of our environment. The environment is man's first right. Without a safe environment, man cannot exist to claim other rights, be they political, social, or economic.

I believe that Shell knows this much and accepts it for the developed world. Unfortunately, in Africa, Shell and many a Western multinational company have failed to apply the environmental principle of "best practice" in their entrepreneurial activities. Using their financial muscle and the political power they exercise over African governments, they are wont to blackmail, terrorize and silence those who have the temerity to question their environmental records. This is not acceptable.

***

I appeal to the international community and environmentalists throughout the world to stand by me and the Ogoni people as we battle to save our devastated environment and for our survival as a people with sustainable development (Saro-Wiwa, 1995).

Saro-Wiwa's struggle is far from won and may never be won. Shell Oil is gone from Ogoniland, and the military dictatorship that executed Saro-Wiwa is gone as well, at least in name, re-placed by a succession of unstable governments. Saro-Wiwa is survived by his father and children. But conditions in the Niger Delta are no better.

Marina Silva (1958–)

The rain forests, biodiversity, and immense expanses of pristine natural areas of the Amazon region have captured the attention and imagination of environmentalists for decades. Mining, urban development, and unsustainable agricultural practices threaten the region; environmentalists within Brazil and abroad seek ways to balance the need to relieve Brazil's intense poverty with the protection of the Amazon. Many indigenous peoples live sustain-ably in the forests; Brazil's rubber-tappers have also found a sus-tainable method of extracting wealth, at least for a small number of people. The rubber-tappers go from tree to tree in the forest, extracting natural latex without harming the trees. Marina Silva grew up as the daughter of a rubber-tapper in Acre, a state in northwestern Brazil. She spent her childhood tapping rubber trees like her father; her health was poor, though, and in 1975, at the age of sixteen, she left the forest for the city to seek medical treatment. She had never learned to read or write, but she worked as a maid while attending school and in 1981 earned a university degree in history from the Federal University of Acre (Hildebrandt, 2001; *Crossing the Divide*, 2004).

At this time, the lifestyle of the rubber-tappers of Amazonia was under attack. Deforestation destroyed the resource upon which rubber-tappers and indigenous Brazilians depended; vio-lent confrontations between settlers and earlier inhabitants of the forest were common. The cause of the rubber-tappers and indige-nous peoples was supported by urban Brazilian and foreign envi-ronmentalists because it coincided with the preservation of the Amazon, but the rubber-tappers themselves were scattered and disorganized. With fellow rubber-tapper Francisco (Chico) Mendes, founder of a rubber-tappers' union, Silva organized a union movement that engaged in *empates*, peaceful demonstra-tions against deforestation. As a result of the *empates*, large areas of forest were set aside as protected natural reserves in which only sustainable extraction, such as rubber-tapping, was permit-ted, but the rubber-tappers' unions attracted the ire of the settlers. Chico Mendes was assassinated in 1988; although he doubtless would have preferred to avoid his martyrdom, his death became a rallying point for environmentalists worldwide, providing added support for the work of Silva and others and ultimately bringing about changes in the Brazilian government's Amazon policy. Two million hectares of forest in Acre are now protected as sustainable extractive reserves.

Marina Silva has not confined her concern for the environment to Acre or to rubber-tapping. In 1994, at the age of thirty-eight, she was elected to the Brazilian Senate—the first rubber-tapper and youngest senator ever elected. In 1996, she was awarded the Goldman Environmental Prize, and in 2003 she was appointed minister of the environment by Brazil's president. As the highest environmental official in one of the world's most critical environmental regions, she is aware of the importance of her task: "Brazil is the world's richest country in terms of biodiversity. We have more than 20 percent of all known living species and in recent years have seen the value of natural resources in creating new products." She is committed to sustainable development, saying:

> . . . the creation of the "socio-environmental" concept, which defends the jungles inseparably from the people who live in them. That struggle seeks to introduce all aspects of sustainable development into the heart of government. Strength grows through combined effort. The Environment Ministry is guided by three pillars: transversality, which is the interaction of environmental policy with all sectors of government; community oversight; and sustainable development (Osava, 2001).

Explaining the unfamiliar concept of "tranversality," she says, "I have always been a critic of the argument that the fight against the destruction of rain forests was the sole responsibility of the Environment Ministry. I think this is a job for the whole government, for all departments" (*Crossing the Divide*, 2004).

Sustainable development may be hard to achieve, however. While Acre's sustainable extractive reserves may preserve the lifestyles of some indigenous peoples and rubber-tappers as well as the forest itself, rubber-tapping as an industry can absorb no more than a few hundred of the tens of millions of desperately poor Brazilians. And rubber-tapping is a hard life; Silva herself did not learn to read or write until she left the forest; her rise from illiterate rubber-tapper to senior government official required extraordinary motivation and ability and is not one that could be easily duplicated. Brazilians must learn to exploit the resources of the forest in other sustainable ways. On the positive side, slash-and-burn agriculture and clear-cutting do not provide luxurious lifestyles, either, thus greatly expanding the potential for finding alternative uses of the forest that will be palatable to settlers.

Even here there are difficulties, however; often the entities most skilled in finding new uses for existing biological resources are foreign corporations, but the activities of these corporations raise bio-piracy concerns. Silva says, "To fight bio-piracy we have the Biodiversity Resource Access Law, which must be reinforced. A recent case of bio-piracy involves the *cupuaçú*, an Amazonian fruit that was patented by a Japanese company" (Osava, 2001). The Japanese company, Asahi Foods, has patented processes for extracting oils from the seeds of the *cupuaçú*; it uses these extracts to make a sweet similar to chocolate, called cupulate. Asahi's *cupuaçú* patents have triggered a storm of protest from international environmental groups, much of it ill-informed—many people seem to believe that Asahi has somehow managed to patent the fruit itself. The adverse publicity may scare off potential investors in sustainable Amazon resource research; Silva will have a difficult job balancing the need for foreign investment against the rights of Amazonian peoples in their knowledge of forest products.

Janos Vargha (1949–) and Duna Kör

Duna Kör (Danube Circle) was, if not the very first, then among the first, grassroots environmental movements in Soviet-dominated Europe. Its emergence coincided with, and to some extent caused, the downfall of Communism in Hungary; it played an important role in the end of the Cold War.

Although proto-environmental associations such as hiking clubs can be traced back at least as far as the twelfth century in Hungary, the imposition of Communist rule after World War II effectively prevented any environmentalist opposition to development and exploitation of Hungary's natural resources. The Communist Party channeled environmentalist sentiment into approved organizations such as the University and College Students' Movement for Environmentalism. Because these organizations were controlled by the Communist Party and because the state, and thus the party, also controlled or at least authorized all development taking place in Hungary it was, as a practical matter, impossible for groups such as the Movement for Environmentalism to oppose any development project or even express their concerns effectively. A few specialized environmentalist groups existed outside the state power structure, including the Magyar Madártani Egyesület (an ornithological society) and Bokor, a Christian religious movement that advocated a simple existence, including refraining from unnecessary consumption and from military service (Regional Environmental Center, 1997).

In 1981, Janos Vargha, a journalist, published an article in a Hungarian magazine criticizing a planned hydroelectric plant on the Danube. Three years later, in 1984, the audience at a lecture by Vargha founded the Committee for the Danube, an ad hoc environmental group dedicated to stopping the hydroelectric project. The committee eventually became the Danube Circle (Duna Kör in Hungarian) (Galambos, 1992).

Duna Kör was not the only group to oppose the project. At least two others, A Dunáért Alapítvány (Foundation for the Danube) and Kékek (the Blues), joined with Duna Kör. This outbreak of environmental opposition presented a problem for an authoritarian government that traditionally suppressed dissent: the environmentalists were not "dissenters" in the traditional sense. At first they did not openly challenge the government's or the party's authority to rule, or even express dissatisfaction with Communist rule. Environmentalist arguments were often phrased in patriotic terms: The hydroelectric project would exploit the resources of Hungary for the benefit of Czechoslovakia and Austria. At the same time, though, the strength of the dissent was due not only to the magnitude of the potential environmental harm from the project, but also to pent-up resentment of decades of one-party rule: Protest against the hydroelectric project could be interpreted in part as protest against Communist rule. As a result, the government retaliated against some prominent protesters by depriving them of their jobs.

In 1985 Duna Kör received the Right Livelihood Award, also known as the Alternative Nobel Prize; the state-controlled Hungarian press suppressed the news of the award. And in 1990, Janos Vargha was awarded the Goldman Environmental Prize. But Duna Kör's finest hour had come a year earlier, in 1989, when the Danube dispute led to the resignation of several members of the government and, far more significantly, free multiparty elections. Even before those elections were held, the members of the Hungarian Parliament responded to the unpopularity of the project and voted to scrap it (Serenyi, 1992; Fitzmaurice, 1998).

But once its initial goal had been achieved, Duna Kör found itself marginalized to some extent. Ties to Slovak environmentalists, in particular, were weakened by Slovakia's independence. The focus of environmentalism has shifted in Hungary. It is no longer the Hungarian government that is the biggest threat to Hungary's environment, but private companies, often from Austria or Germany. Janos Vargha has become the president of Ister, an environmental research institute.

Sources and Further Reading

Beanal v. Freeport-McMoRan, Inc. 969 F. Supp. 362 (E.D. La. 1997), aff'd F. 3d, 1975th Cir. 161 (1999).

Bongaarts, John. *Population: Ignoring Its Impact*, Scientific American. Available at http://www.sciam.com/article.cfm?articleID=000F3D47-C6D2–1CEB-93F6809EC5880000&pageNumber=7&catID=2, January, 2002 (visited 17 September 2004).

Borneo Resources Institute. http://brimas.www1.50megs.com/ (visited September 17, 2004).

Brundtland, Gro Harlem. *Madam Prime Minister: A Life in Power and Politics.* New York: Farrar, Straus and Giroux, 2002.

Cardwell, Peter. *Hot Air Or Real Scare? Are Initiatives like Kyoto Worth the Effort?* Oxford Student. Available at http://www.oxfordstudent.com/2004–06–03/features/3, 3 June 2004 (visited 12 September 2004).

Crossing the Divide, Part 2: Out of the Amazon. Available at http://www.tve.org/earthreport/archive/doc.cfm?aid=1520 (visited 18 September 2004).

Cupuaçú Case. Available at http://www.amazonlink.org/biopiracy/cupuacu.htm (visited 18 September 2004).

Domoto, Akiko, and Kunio Iwatsuki, eds. *Threat to Life: The Impact of Climate on Japan's Biodiversity.* Washington, DC: Island Press, 2000.

Ege, Christian, and Jeanne Lind Christiansen, eds. *Sceptical Questions and Sustainable Answers.* Kirsten Trolle-Hansen, trans. Copenhagen: Danish Ecological Council, 2002.

Feshbach, Murray, and Alfred Friendly, Jr. *Ecocide in the USSR: Health and Nature Under Siege.* New York: Basic Books, 1992.

Fitzmaurice, John. *Damming the Danube: Gabčíkovo and Post-Communist Politics in Europe.* Boulder, CO: Westview Press, 1998.

The Flames of Shell: Oil, Nigeria and the Ogoni. Available at Third World Traveler, http://www.thirdworldtraveler.com/Boycotts/Flames_Shell.html (visited 17 September 2004).

Fog, Kåre. *Lomborg Errors.* Available at http://www.lomborg-errors.dk/ (visited 17 September 2004).

French, Hilary F. *Green Revolutions: Environmental Reconstruction in Eastern Europe and the Soviet Union.* Washington, DC: Worldwatch Paper No. 99, 1990.

Gabčíkovo -Nagymaros Dispute (Slovakia v. Hungary), 1997. I.C.J. 7 (1997).

Galambos, Judit. *Political Aspects of an Environmental Conflict: The Case of the Gabčíkovo-Nagymaros Dam System.* Perspectives of Environmental Conflict and International Relations, 75–76 (Jyrki Kakonen, ed., 1992).

Gedicks, Al, and Roger Moody. *Resource Rebels.* Cambridge, MA: South End Press, 2001.

Gelb, Joyce. *Gender Policies in Japan and the United States: Comparing Women's Movements, Rights, and Politics.* Houndmills, Basingstoke, UK: Palgrave Macmillan, 2003.

Goldman Environmental Prize: Harrison Ngau Laing, 1990. Available at http://www.goldmanprize.org/recipients/recipient Profile.cfm?recipientID=37 (visited 17 September 2004).

Goldman Environmental Prize: Vera Mischenko, 2000. Available at http://www.goldmanprize.org/recipients/recipients.html.

Greenberg, Jonathan R., and Ken Saro-Wiwa. *African Postcolonial Literature in English in the Postcolonial Web, 1990.* Available at http://65.107.211.208/sarowiwa/sarowiwabio.html (visited 10 September 2004).

Greenpeace. *First Anniversary of the Death of Ken Saro-Wiwa, 10 November 1996.* Available at http://archive.greenpeace.org/comms/ken/ (visited 17 September 2004).

Hildebrandt, Ziporah. *Marina Silva: Defending Rainforest Communities in Brazil.* New York: Feminist Press, 2001.

Holdren, John P. *Energy: Asking the Wrong Question.* Scientific American, January 2002. Available at http://www.sciam.com/article.cfm?articleID=000F3D47-C6D2–1CEB 93F6809EC5880000&pageNumber=5& catID=2 (visited 17 September 2004).

Human Rights Watch. *Indonesia: Human Rights and Pro-Independence Actions in Papua, 1999–2000,* May 2000. Available at http://www.hrw.org/reports/2000/papua/index.htm#Top*OfPage* (visited 12 September 2004).

Iban and Bidayuh Communities Suing Oil Palm Company for Encroachment of NCR. Available at http://brimas.www1.50megs.com/PR-16–9–02.htm (visited 17 September 2004).

Jedrzej, George, and Scott Pegg, eds. *Transnational Corporations and Human Rights.* Houndmills, Basingstoke, UK: Palgrave Macmillan, 2003.

Kagohashi, Taka'aki. *The Amami Rights of Nature Lawsuit.* Social Science Japan, 14, April 2002.

Kingsbury, Damien, and Harry Aveling, eds. *Autonomy and Disintegration in Indonesia.* London: Routledge Curzon, 2003.

Lomborg, Bjørn. *Institut for Miljøvurdering Press Release, 12 March 2004.* Available at www.lomborg.com/files/Scientific%20Dishonesty%20case %20closed%20-%20Lomborg%20cleared.pdf (visited 8 September 2004).

Lomborg, Bjørn. *The Skeptical Environmentalist: Measuring the Real State of the World.* Cambridge, UK: Cambridge University Press, 2001.

———. *Global Crises, Global Solutions.* Cambridge, UK: Cambridge University Press, 2004.

Lovejoy, Thomas. *Biodiversity: Dismissing Scientific Process.* Scientific American, January 2002. Available at http://www.sciam.com/article. cfm?articleID=000F3D47-C6D2–1CEB-93F6809EC5880000&page Number=10&catID=2 (visited 17 September 2004).

Lynas, Mark. *High Tide: The Truth About Our Climate Crisis.* New York: Picador, 2004.

Mischenko, Vera, and Erika Rosenthal. *Russian Environmentalists Sue President Putin Over Dissolution of Environmental Regulatory Agency: Citizens Petition Supreme Court to Uphold the Rule of Law,* Earthjustice Newsroom, 22 August 2000. Available at http://www.earthjustice.org/news/ display.html?ID=140 (visited 19 September 2004).

Murphy, Dan. *Violence, a U.S. Mining Giant, and Papua Politics.* Christian Science Monitor, 1, 3 September 2002.

Osava, Mario. *The Anthill from Below and from Above,* TierrAmérica: Medio Ambiente y Desarollo, 2001. Available at http://www.tierramerica.org/2003/0414/idialogos.shtml (visited 18 September 2004) [in English].

Pimm, Stuart, and Jeff Harvey. *No Need to Worry about the Future: Environmentally, We Are Told, 'Things Are Getting Better.'* Nature, November 2001. Available at http://www.nature.com/cgi.taf/DynaPage.taf?file=/nature/journal/v414/n6860/index.html (visited 17 September 2004) (payment is required).

President, GLOBE Japan. Available at http://www.globeinternational. org/archives/earthsummit/earth5rio-biodiversity.html.

"Regional Environmental Center, Country Report: Hungary," in *Problems, Progress and Possibilities: A Needs Assessment of Environmental NGOs in Central and Eastern Europe*, April 1997. Available at http://www.rec. org/REC/Publications/NGONeeds/Hungary.html (visited 14 September 2004).

Rennie, John. *Misleading Math about the Earth: Science Defends Itself against The Skeptical Environmentalist*. Scientific American, January 2002. Available at http://www.sciam.com/article.cfm?articleID=000F3D47-C6D2–1CEB-93F6809EC5880000&pageNumber=1&catID=2.

Rio+5 Forum: Report on Biodiversity. Submitted by Akiko Domoto.

Rosaldo, Renato. *Cultural Citizenship in Island Southeast Asia: Nation and Belonging in the Hinterlands*. Berkeley: University of California Press, 2003.

Santana, Rebecca. *In Oil We Trust? Russian Islanders Have Mixed Feelings: Billions in Western Projects Provide Jobs but also Create Eco-fears*, MSNBC News, 13 October 2003. Available at http://www.msnbc.msn.com/id/3158398/ (visited 19 September 2004).

Sarawak Online. Available at http://www.sarawak.gov.my/contents/population/population.shtml (visited 17 September 2004).

Saro-Wiwa, Ken. *Genocide in Nigeria: The Ogoni Tragedy*. Port Harcourt, NGA: Saros International Publishers, 1992.

———. *Songs in a Time of War*. London: Lynne Rienner Publishers, 1985.

———. *Sozaboy, 1986*. London: Longman African Writers, 1985.

———. *Stand by Me and the Ogoni People*. Earth Island Journal, 1995. Available at http://www.earthisland.org/journal/ogoni.html (visited 18 September 2004).

Schneider, Stephen. *Global Warming: Neglecting the Complexities*, Scientific American, January 2002. Available at http://www.sciam.com/article. cfm?articleID=000F3D47-C6D2–1CEB-93F6809EC5880000&pageNumber=1&catID=2.

Schwabach, Aaron. *Diverting the Danube: The Gabčíkovo-Nagymaros Dispute and International Freshwater Law*. 14 Berkeley Journal of International Law 290 (1996).

Serenyi, Juliet. *Danube Project Sours*. Christian Science Monitor, 19, 9 December 1992.

Shellman Says Sorry. The Economist, 8 May 1997.

Silva, Senator Marina. Available at http://www.senado.gov.br/web/senador/marinasi/marinasi.htm (visited 18 September 2004) [in Portuguese].

The Stars of Asia—Policymakers: Akiko Domoto, Governor, Chiba Prefecture, Japan, Business Week Online, 2 July 2001. Available at http://www. businessweek.com/magazine/content/01_27/b3739046.htm (visited 19 September 2004).

Tom Beanal's Speech at Loyola University, May 23, 1996 (Translated from Bahasa). Available on many websites, including http://www. corpwatch.org/article.php?id=987 (visited 30 May 2004).

UCS Examines The Skeptical Environmentalist by Bjørn Lomborg. Union of Concerned Scientists, December 2003. Available at http://www.ucsusa. org/global_environment/archive/page.cfm?pageID=533 (visited Sept. 10, 2004).

Vargha, Janos. *Egyre tavolabb a jotol.* Valosag, November 1981 (in Hungarian).

Watts, Michael. "Black Gold, White Heat." In Geographies of Resistance (Steve Pile and Michael Keith eds., Routledge, 1997).

Wiwa, Ken. *In the Shadow of a Saint: A Son's Journey to Understand His Father's Legacy.* Hanover, NH: Steerforth Press, 2001. [Ken Wiwa is Ken Saro-Wiwa's son.]

World Commission on Environment and Development. *Our Common Future.* Oxford: Oxford University Press, 1987.

World Health Organization. *Dr. Gro Harlem Brundtland, Director-General* (capsule biography). Available at http://www.who.int/dg/bruntland/ en/ (visited 19 September 2004).

www.anti-lomborg.com, http://www.mylinkspage.com/lomborg.html (visited 17 September 2004).

6

Treaties, Cases, Reports, and Other Documents

The selection of documents in this chapter is necessarily representative rather than encyclopedic. The document excerpts reproduced below provide an overview of some of the highlights of international environmental law; those doing research in the field, however, will want to locate the full texts of many additional documents. Most of the cases, treaties, United Nations documents, and other materials produced by governmental and intergovernmental organizations are easily available on the Web. The bibliographic materials in Chapter 8 include Web addresses where available, and many of the organizations listed in Chapter 7 provide online libraries of documents within their area of specialization. In addition, most of these documents can be located by typing the title of the document into the search bar of a search engine such as Google.

Treaties and laws are generally available for free. Materials protected by copyright, such as the *Restatement (Third) of Foreign Relations of the United States,* are available online via paid proprietary services such as Westlaw (www.westlaw.com); a fee is charged for access. They may also be purchased from booksellers and are available in many public libraries.

Sources of International Law

Statute of the International Court of Justice, Articles 38 and 59

The International Court of Justice is the judicial organ of the United Nations; its functions, powers, and jurisdiction are defined by its

185

statute. Article 38 of the statute sets out the sources for the court to use in determining the rules of international law applicable to cases brought before it. That is all that Article 38 does; it is a list of sources of authority for the court, and nothing more. But in the absence of any universally accepted statement of the sources of international law, the list of sources in Article 38(1) is the closest thing to a universally accepted starting point. It is just that, however: a starting point, not an ending point.

Article 38

1. The Court, whose function is to decide in accordance with international law such disputes as are submitted to it, shall apply:

a. international conventions, whether general or particular, establishing rules expressly recognized by the contesting states;

b. international custom, as evidence of a general practice accepted as law;

c. the general principles of law recognized by civilized nations;

d. subject to the provisions of Article 59, judicial decisions and the teachings of the most highly qualified publicists of the various nations, as subsidiary means for the determination of rules of law.

2. This provision shall not prejudice the power of the Court to decide a case ex aequo et bono, if the parties agree thereto.

***

Article 59   The decision of the Court has no binding force except between the parties and in respect of that particular case.

Source: 1976 Y.B.U.N. 1052, 59 Stat. 1031, T.S. No. 993

The Restatement (Third) of the Foreign Relations Law of the United States, §§102–03

The Restatement (Third) of the Foreign Relations Law of the United States is a text compiled and published by the American Law Institute, an Nongovernmental Organization (NGO). The ALI publishes Restate-

ments of many areas of U.S. law, from Agency to Unfair Competition. The Restatements are designed to provide a comprehensive overview of the law in a particular area—in the ALI's own words, to "tell judges and lawyers what the law is." The ALI itself is not a law-making body, however, and although the Restatements are widely respected, they have no independent legal effect. They may incorporate the errors, aspirations, or prejudices of the legal scholars who compile them. With foreign relations law, a highly politicized subject, the actual practice of the United States may vary widely from one administration to the next and even over the course of a single administration. It is interesting to compare the Restatement's list of the sources of international law to the list in Article 38(1) of the Statute of the International Court of Justice.

Introductory Note

International law is the law of the international community of states. It deals with the conduct of nation-states and their relations with other states, and to some extent also with their relations with individuals, business organizations, and other legal entities . . . .

In principle, law that has been generally accepted cannot be later modified unilaterally by any state . . ., but particular states and groups of states can contribute to the process of developing (and modifying) law by their actions as well as by organized attempts to achieve formal change.

Custom and international agreement. International law is made in two principal ways—by the practice of states ("customary law") and by purposeful agreement among states (sometimes called "conventional law")[.]

***

§102. Sources Of International Law

(1) A rule of international law is one that has been accepted as such by the international community of states

(a) In the form of customary law;

(b) By international agreement; or

(c) By derivation from general principles common to the major legal systems of the world.

(2) Customary international law results from a general and consistent practice of states followed by them from a sense of legal obligation.

(3) International agreements create law for the states parties thereto and may lead to the creation of customary international

law when such agreements are intended for adherence by states generally and are in fact widely accepted.

(4) General principles common to the major legal systems, even if not incorporated or reflected in customary law or international agreement, may be invoked as supplementary rules of international law where appropriate.

Comment:

a. *Sources and evidence of international law distinguished.* This section indicates the ways in which rules or principles become international law. The means for proving that a rule or principle has in fact become international law in one of the ways indicated in this section is dealt with in §103.

b. *Practice as customary law.* "Practice of states," Subsection (2), includes diplomatic acts and instructions as well as public measures and other governmental acts and official statements of policy. . . . Failure of a significant number of important states to adopt a practice can prevent a principle from becoming general customary law though it might become "particular customary law" for the participating states. See Comment e. A principle of customary law is not binding on a state that declares its dissent from the principle during its development. See Comment d.

c. *Opinio juris.* For a practice of states to become a rule of customary international law it must appear that the states follow the practice from a sense of legal obligation (opinio juris sive necessitatis); a practice that is generally followed but which states feel legally free to disregard does not contribute to customary law. A practice initially followed by states as a matter of courtesy or habit may become law when states generally come to believe that they are under a legal obligation to comply with it. It is often difficult to determine when that transformation into law has taken place. Explicit evidence of a sense of legal obligation (e.g., by official statements) is not necessary; opinio juris may be inferred from acts or omissions.

d. *Dissenting views.* Although customary law may be built by the acquiescence as well as by the actions of states (Comment b) and become generally binding on all states, in principle a state that indicates its dissent from a practice while the law is still in the process of development is not bound by that rule even after it matures[.]

e. *General and special custom.* The practice of states in a regional or other special grouping may create "regional," "special," or "particular" customary law for those states [among each other].

***

k. *Peremptory norms of international law (jus cogens).* Some rules of international law are recognized by the international community of states as peremptory, permitting no derogation. These rules prevail over and invalidate international agreements and other rules of international law in conflict with them[.]

l. *General principles as secondary source of law.* Much of international law, whether customary or constituted by agreement, reflects principles analogous to those found in the major legal systems of the world, and historically may derive from them or from a more remote common origin . . . . General principles common to systems of national law may be resorted to as an independent source of law. That source of law may be important when there has not been practice by states sufficient to give the particular principle status as customary law and the principle has not been legislated by general international agreement.

General principles are a secondary source of international law, resorted to for developing international law interstitially in special circumstances.

***

Reporters' Notes

1. *Statute of International Court of Justice and sources of law.* This section draws on Article 38(1) of the Statute of the International Court of Justice. . . .

The statute . . . does not use the term "sources," but this restatement follows common usage in characterizing customary law, international agreements, and general principles of law as "sources" of international law, in the sense that they are the ways in which rules become, or become accepted as, international law. International lawyers sometimes also describe as "sources" the "judicial decisions and the teachings of the most highly qualified publicists of the various nations," mentioned in Article 38(1)(d) of the Statute of the Court, supra. Those, however, are not sources in the same sense since they are not ways in which law is made or accepted, but opinion-evidence as to whether some rule has in fact become or been accepted as international law. See §103.

***

§103. Evidence of International Law

(1) Whether a rule has become international law is determined by evidence appropriate to the particular source from which that rule is alleged to derive (§102).

(2) In determining whether a rule has become international law, substantial weight is accorded to

(a) Judgments and opinions of international judicial and arbitral tribunals;

(b) Judgments and opinions of national judicial tribunals;

(c) The writings of scholars;

(d) Pronouncements by states that undertake to state a rule of international law, when such pronouncements are not seriously challenged by other states.

Source: *The Restatement (Third) of the Foreign Relations Law of the United States,* §§102–03 (Philadelphia: American Law Institute, 1987)

Treaties

Antarctic Treaty

The Antarctic Treaty was originally designed to address the territorial dispute over Antarctica, rather than as an environmental protection treaty. Article V of the treaty contained environmental measures: a prohibition on nuclear explosions and the disposal of radioactive waste in Antarctica. Over the years since 1959, when the Antarctic Treaty was concluded, there has been a shift in attitudes toward and perceptions of the continent. Antarctica is no longer valued for its territory, which is for all practical purposes uninhabitable, or for its military significance, which is essentially nil. Instead, it is valued for the unique opportunities it offers for scientific research, for its potential mineral resources, and for its pristine natural environment. The tension between the first two of these and the third has dominated subsequent international rule-making. Possible conflicts between scientific research and environmental protection surfaced in the case of Environmental Defense Fund v. Massey, *an excerpt from which is reproduced later in this chapter. And the possibility of commercial exploitation of Antarctica's mineral resources led to the unsuccessful* attempt during the 1980s to bring into force an Antarctic Mineral Resources Convention, and the later, ultimately successful, Antarctic Environment Protocol. This was made possible in part by the quasi-governmental structure set up by Article IX of the Antarctic Treaty, which charged the parties with, among other tasks, considering and recommending measures for the "preservation and conservation of living resources in Antarctica."

Article V

1. Any nuclear explosions in Antarctica and the disposal there of radioactive waste material shall be prohibited.

2. In the event of the conclusion of international agreements concerning the use of nuclear energy, including nuclear explosions and the disposal of radioactive waste material, to which all of the Contracting Parties whose representatives are entitled to participate in the meetings provided for under Article IX are parties, the rules established under such agreements shall apply in Antarctica.

***

Article IX

1. Representatives of the Contracting Parties named in the preamble to the present Treaty shall meet at the City of Canberra within two months after date of entry into force of the Treaty, and thereafter at suitable intervals and places, for the purpose of exchanging information, consulting together on matters of common interest pertaining to Antarctica, and formulating and considering, and recommending to their Governments, measures in furtherance of the principles and objectives of the Treaty including measures regarding:

(a) Use of Antarctica for peaceful purposes only;

(b) Facilitation of scientific research in Antarctica;

(c) Facilitation of international scientific cooperation in Antarctica;

(d) Facilitation of the exercise of the rights of inspection provided for in Article VII of the Treaty;

(e) Questions relating to the exercise of jurisdiction in Antarctica;

(f) Preservation and conservation of living resources in Antarctica.

Source: Antarctic Treaty, Dec. 1, 1959, 12 U.S.T. 794, 402 U.N.T.S. 71.

Antarctic Treaty Environmental Protection Protocol

By the 1980s, Antarctica had, with the help of innumerable television specials, magazine articles, coffee table books, and the like, been firmly established in the popular imagination as Earth's last untouched continent: A land of Emperor penguin rookeries stretching to the horizon, of blue whales feeding on immense clouds of krill, of Weddell seals diving half a mile beneath the surface of the ocean, and ferocious, gigantic Leopard seals, a dazzling white landscape of beautiful, if stark, purity. Most were probably content to leave Antarctica as a sort of global park, although few would ever have the opportunity to visit it; the idea of sacrificing Antarctica's wilderness to lower the price of oil by thirty cents a barrel was in some way more disturbing than the possibly more destructive consequences of oil drilling in long-populated areas such as the Persian Gulf.

By 1988, the parties to the Antarctic Treaty had drafted and signed the Antarctic Mineral Resources Convention, which provided rather weak protections. Two of the signing parties, Australia and France, were so dismayed by the weakness of the Mineral Resources Convention that they subsequently refused to ratify it. The convention never came into force; instead, the Consultative Parties to the Antarctic Treaty created a protocol to the original treaty, formally establishing Antarctica as a "natural reserve, devoted to peace and science." The Antarctic Environment Protocol established a fifty-year moratorium on all mineral extraction activities in Antarctica, other than minor extractions undertaken for purposes of scientific research.

The two purposes enshrined in the protocol may still come into conflict with each other, however. Not all scientific research is environmentally benign; in addition, scientists need to eat, sleep, and move around, and those activities have environmental impacts. In Environmental Defense Fund v. Massey, *excerpted below, the court applied a U.S. law requiring environmental impact assessment to activities by a U.S. government agency in Antarctica. In 1993, at the time the case was decided, the Antarctic Environment Protocol had not yet entered into force; the United States had ratified the protocol in 1991, but it did not enter into force until 1998. It is possible that, had the protocol entered into force earlier, the outcome in Massey might have been different.*

Article 8: Environmental Impact Assessment

1. Proposed activities referred to in paragraph 2 below shall be subject to the procedures set out in Annex I for prior assessment of the impacts of those activities on the Antarctic environ-

ment or on dependent or associated ecosystems according to whether those activities are identified as having:

(a) less than a minor or transitory impact;

(b) a minor or transitory impact; or

(c) more than a minor or transitory impact.

2. Each Party shall ensure that the assessment procedures set out in Annex I are applied in the planning processes leading to decisions about any activities undertaken in the Antarctic Treaty area pursuant to scientific research programmes, tourism and all other governmental and non-governmental activities in the Antarctic Treaty area for which advance notice is required under Article VII(5) of the Antarctic Treaty, including associated logistic support activities.

3. The assessment procedures set out in Annex I shall apply to any change in an activity whether the change arises from an increase or decrease in the intensity of an existing activity, from the addition of an activity, the decommissioning of a facility, or otherwise.

4. Where activities are planned jointly by more than one Party, the Parties involved shall nominate one of their number to coordinate the implementation of the environmental impact assessment procedures set out in Annex I.

Source: *Protocol on Environmental Protection to the Antarctic Treaty*, 30 I.L.M. 1461, Oct. 4, 1991.

Basel and Bamako Conventions

Article 4 of the Basel Convention was heavily criticized by the people and governments of developing countries, especially African countries. Article 4 was perceived as licensing the export of hazardous waste to these countries. This perceived callousness was particularly evident in Article 4.6, which prevents "the export of hazardous wastes or other wastes for disposal within the area south of 60° South latitude." Some took this to mean that the countries of the developed world deemed the penguins and seals of uninhabited Antarctica more worthy of protection than the human beings of Africa, although in retrospect the slight appears to have been unintended.

As a result, several African countries united to form the Bamako Convention, prohibiting the import of hazardous wastes into the territory of member states from outside those states. The parties to the Basel Convention subsequently attempted to ban such trade themselves, with inconclusive results: In 1994, the second meeting of the Basel Convention's COP (COP-2) agreed to ban the export of hazardous wastes from Organization for Economic Cooperation and Development (OECD) to non-OECD countries. The ban on export of wastes for final disposal was to be effective immediately, and a ban on export for recovery and recycling was to take effect by the end of 1997. The process by which this decision was reached did not comply with the requirements of Article 17 regarding amendment of the convention, and thus was not incorporated into the text of the Basel Convention and may not have been legally binding. In 1995 the third meeting of the COP proposed that the ban be incorporated as a formal amendment, in somewhat modified form: The earlier reference to OECD and non-OECD countries was replaced with a reference to Annex VII countries (parties to the Basel Convention that are also members of the OECD and/or the EU, plus Liechtenstein) and non-Annex VII countries (all other parties to the convention). In order to enter into force, the Ban Amendment (as it is now known) must be ratified by three-fourths of the parties present at the time of its adoption; it has not yet entered into force and may never do so.

Note that although Article 4.5 might appear to exclude imports or exports of hazardous wastes between Basel Convention states and the United States (a non-party), Article 11 creates a loophole allowing such shipments if they "do not derogate from the environmentally sound management of hazardous wastes and other wastes as required by" the convention and are communicated to the secretariat.

Basel Convention
Article 4: General Obligations

1. (a) Parties exercising their right to prohibit the import of hazardous wastes or other wastes for disposal shall inform the other Parties of their decision pursuant to Article 13.

(b) Parties shall prohibit or shall not permit the export of hazardous wastes and other wastes to the Parties which have prohibited the import of such wastes, when notified pursuant to subparagraph (a) above.

(c) Parties shall prohibit or shall not permit the export of hazardous wastes and other wastes if the State of import does not consent in writing to the specific import, in the case where that State of import has not prohibited the import of such wastes.

2. Each Party shall take the appropriate measures to:

(a) Ensure that the generation of hazardous wastes and other wastes within it is reduced to a minimum, taking into account social, technological and economic aspects;

(b) Ensure the availability of adequate disposal facilities, for the environmentally sound management of hazardous wastes and other wastes, that shall be located, to the extent possible, within it, whatever the place of their disposal;

(c) Ensure that persons involved in the management of hazardous wastes or other wastes within it take such steps as are necessary to prevent pollution due to hazardous wastes and other wastes arising from such management and, if such pollution occurs, to minimize the consequences thereof for human health and the environment;

(d) Ensure that the transboundary movement of hazardous wastes and other wastes is reduced to the minimum consistent with the environmentally sound and efficient management of such wastes, and is conducted in a manner which will protect human health and the environment against the adverse effects which may result from such movement;

(e) Not allow the export of hazardous wastes or other wastes to a State or group of States belonging to an economic and/or political integration organization that are Parties, particularly developing countries, which have prohibited by their legislation all imports, or if it has reason to believe that the wastes in question will not be managed in an environmentally sound manner, according to criteria to be decided on by the Parties at their first meeting;

**

3. The Parties consider that illegal traffic in hazardous wastes or other wastes is criminal.

**

5. A Party shall not permit hazardous wastes or other wastes to be exported to a non-Party or to be imported from a non-Party.

6. The Parties agree not to allow the export of hazardous wastes or other wastes for disposal within the area south of 60° South latitude, whether or not such wastes are subject to transboundary movement.

**

9. Parties shall take the appropriate measures to ensure that the transboundary movement of hazardous wastes and other wastes only be allowed if:

(a) The State of export does not have the technical capacity and the necessary facilities, capacity or suitable disposal sites in order to dispose of the wastes in question in an environmentally sound and efficient manner; or

(b) The wastes in question are required as a raw material for recycling or recovery industries in the State of import; or

(c) The transboundary movement in question is in accordance with other criteria to be decided by the Parties, provided those criteria do not differ from the objectives of this Convention.

10. The obligation under this Convention of States in which hazardous wastes and other wastes are generated to require that those wastes are managed in an environmentally sound manner may not under any circumstances be transferred to the States of import or transit.

11. Nothing in this Convention shall prevent a Party from imposing additional requirements that are consistent with the provisions of this Convention, and are in accordance with the rules of international law, in order better to protect human health and the environment.

***

13. Parties shall undertake to review periodically the possibilities for the reduction of the amount and/or the pollution potential of hazardous wastes and other wastes which are exported to other States, in particular to developing countries.

***

Article 11: Bilateral, Multilateral, and Regional Agreements

1. Notwithstanding the provisions of Article 4, paragraph 5, Parties may enter into bilateral, multilateral, or regional agreements or arrangements regarding transboundary movement of hazardous wastes or other wastes with Parties or non-Parties provided that such agreements or arrangements do not derogate from the environmentally sound management of hazardous wastes and other wastes as required by this Convention. These agreements or arrangements shall stipulate provisions which are not less environmentally sound than those provided for by this Convention in particular taking into account the interests of developing countries.

2. Parties shall notify the Secretariat of any bilateral, multilateral or regional agreements or arrangements referred to in

paragraph 1 and those which they have entered into prior to the entry into force of this Convention for them, for the purpose of controlling transboundary movements of hazardous wastes and other wastes which take place entirely among the Parties to such agreements. The provisions of this Convention shall not affect transboundary movements which take place pursuant to such agreements provided that such agreements are compatible with the environmentally sound management of hazardous wastes and other wastes as required by this Convention.

***

Article 17: Amendment of the Convention

1. Any Party may propose amendments to this Convention and any Party to a protocol may propose amendments to that protocol. Such amendments shall take due account, inter alia, of relevant scientific and technical considerations.

2. Amendments to this Convention shall be adopted at a meeting of the Conference of the Parties. Amendments to any protocol shall be adopted at a meeting of the Parties to the protocol in question. The text of any proposed amendment to this Convention or to any protocol, except as may otherwise be provided in such protocol, shall be communicated to the Parties by the Secretariat at least six months before the meeting at which it is proposed for adoption. The Secretariat shall also communicate proposed amendments to the Signatories to this Convention for information.

3. The Parties shall make every effort to reach agreement on any proposed amendment to this Convention by consensus. If all efforts at consensus have been exhausted, and no agreement reached, the amendment shall as a last resort be adopted by a three-fourths majority vote of the Parties present and voting at the meeting, and shall be submitted by the Depositary to all Parties for ratification, approval, formal confirmation or acceptance.

4. The procedure mentioned in paragraph 3 above shall apply to amendments to any protocol, except that a two-thirds majority of the Parties to that protocol present and voting at the meeting shall suffice for their adoption.

5. Instruments of ratification, approval, formal confirmation or acceptance of amendments shall be deposited with the Depositary. Amendments adopted in accordance with paragraphs 3 or 4 above shall enter into force between Parties having

accepted them on the ninetieth day after the receipt by the Depositary of their instrument of ratification, approval, formal confirmation or acceptance by at least three-fourths of the Parties who accepted them or by at least two thirds of the Parties to the protocol concerned who accepted them, except as may otherwise be provided in such protocol. The amendments shall enter into force for any other Party on the ninetieth day after that Party deposits its instrument of ratification, approval, formal confirmation or acceptance of the amendments.

6. For the purpose of this Article, "Parties present and voting" means Parties present and casting an affirmative or negative vote.

The proposed Ban Amendment, not yet in force, would add a new paragraph to the preamble and a new Article 4A to the body of the Convention:

Insert new preambular paragraph 7 bis:

Recognizing that transboundary movements of hazardous wastes, especially to developing countries, have a high risk of not constituting an environmentally sound management of hazardous wastes as required by this Convention;

Insert new Article 4A:

1. Each Party listed in Annex VII shall prohibit all transboundary movements of hazardous wastes which are destined for operations according to Annex IV A, to States not listed in Annex VII.

2. Each Party listed in Annex VII shall phase out by 31 December 1997, and prohibit as of that date, all transboundary movements of hazardous wastes under Article 1, paragraph 1 (a) of the Convention which are destined for operations according to Annex IV B to States not listed in Annex VII. Such transboundary movements shall not be prohibited unless the wastes in question are characterized as hazardous under the Convention.

Although it originated as a reaction to inadequacies in the Basel Convention, the Bamako Convention's provisions are for the most part similar to those of the Basel Convention. In Article 4, however, particularly paragraphs 4(1), 4(3)(b), 4(3)(f), and 4(3)(g), the differences are quite pronounced:

Article 4: General Obligations

1. Hazardous Waste Import Ban

All Parties shall take appropriate legal, administrative and other measures within the area under their jurisdiction to prohibit the import of all hazardous wastes, for any reason, into Africa from non-Contracting Parties. Such import shall be deemed illegal and a criminal act. All Parties shall:

(a) Forward as soon as possible, all information relating to such illegal hazardous waste import activity to the Secretariat who shall distribute the information to all Contracting Parties;

(b) Co-operate to ensure that no imports of hazardous wastes from a non-Party enter a Party to this Convention. To this end, the Parties shall, at the Conference of the Contracting Parties, consider other enforcement mechanisms.

***

3. Waste Generation in Africa

Each Party Shall:

***

(b) Impose strict, unlimited liability as well as joint and several liability on hazardous waste generators;

***

The Adoption of Precautionary Measures:

(f) Each Party shall strive to adopt and implement the preventive, precautionary approach to pollution problems which entails, *inter-alia*, preventing the release into the environment of substances which may cause harm to humans or the environment without waiting for scientific proof regarding such harm. The Parties shall co-operate with each other in taking the appropriate measures to implement the precautionary principle to pollution prevention through the application of clean production methods, rather than the pursuit of a permissible emissions approach based on assimilative capacity assumptions;

(g) In this respect Parties shall promote clean production methods applicable to entire product life cycles including: raw material selection, extraction and processing; product conceptualisation, design, manufacture and assemblage; materials transport during all phases; industrial and household usage; reintroduction of the product into industrial systems or nature when it no longer serves a useful function[.] Clean production shall not include "end-of-pipe" pollution controls such as filters and scrubbers, or chemical, physical or biological treatment. Measures which reduce the volume of waste by incineration or concentration, mask the hazard by dilution, or transfer pollutants from one environmental medium to another, are also excluded[.]

Sources:

Basel Convention on the Control of Transboundary Movement of Hazardous Wastes and their Disposal, Mar. 22, 1989, 28 I.L.M. 657.

Basel Convention COP-3, Decision III/1: Amendment to the Basel Convention, available at http://www.basel.int/pub/baselban.html (visited September 1, 2004).

Bamako Convention on the Ban of Imports into Africa and the Control of Transboundary Movement and Management of Hazardous Wastes within Africa, Jan. 30, 1991, 30 I.L.M. 773.

ENMOD

During the Vietnam War, the United States deliberately destroyed huge tracts of forest in order to prevent enemy forces from using them as cover. While some military advantage was gained, the magnitude of the environmental harm was immense. Deliberate environmental destruction for military purposes was not new; for example, nearly a century earlier the American bison had been hunted nearly to extinction in a deliberate and successful effort to destroy the traditional lifestyle of the Plains Indians. But advances in technology had greatly increased the potential for such environmental harm. To reduce the risk of massive environmental disaster being created to achieve some minor or transient military objective, two treaties were concluded: ENMOD and Protocol I to the Geneva Conventions. The two contain superficially similar definitions of the harm to be avoided, but the choice of conjunctions—"or" in ENMOD and "and" in Protocol I—has a dramatic legal effect; ENMOD covers a broader range of activities and environmental harms. Note that the Rome Statute of the International Criminal Court also uses "and."

Article I

1. Each State Party to this Convention undertakes not to engage in military or any other hostile use of environmental modification techniques having widespread, long-lasting or severe effects as the means of destruction, damage or injury to any other State Party.

2. Each State Party to this Convention undertakes not to assist, encourage or induce any State, group of States or international organization to engage in activities contrary to the provisions of paragraph 1 of this article.

Article II

As used in Article I, the term "environmental modification techniques" refers to any technique for changing—through the deliberate manipulation of natural processes—the dynamics, composition or structure of the Earth, including its biota, lithosphere, hydrosphere and atmosphere, or of outer space.

Source: *Convention on the Prohibition of Military or Any Other Hostile Use of Environmental Modification Techniques*, Dec. 10, 1976, 31 U.S.T. 333, 1108 U.N.T.S. 151.

Kuwait Regional Convention for Co-operation on the Protection of the Marine Environment from Pollution

The ocean is protected from pollution by a complex network of treaties. The UN Convention on the Law of the Sea provides some overarching provisions, and a section dealing with the specific problem of deep seabed mining. Other global treaties address other specific problems. But because regional seas often have regional problems, several regional seas treaties protect specific bodies of water. The Kuwait Convention reproduced below is one such treaty; it protects the Persian Gulf. The greatest environmental threat to the Gulf is the oil industry; as can be seen from the excerpt below, the Kuwait Convention contains specific provisions to address that threat.

Article III: General Obligations

(a) The Contracting States shall, individually and/or jointly, take all appropriate measures in accordance with the present Convention and those protocols in force to which they are party to prevent, abate and combat pollution of the marine environment in the Sea Area;

(b) In addition to the Protocol concerning Regional Co-operation in Combating Pollution by Oil and other Harmful Substances in Cases of Emergency opened for signature at the same time as the present Convention, the Contracting States shall co-operate in the formulation and adoption of other protocols prescribing agreed measures, procedure and standards for the implementation of the Convention;

(c) The Contracting States shall establish national standards, laws and regulations as required for the effective discharge of the obligation prescribed in paragraph (a) of this article, and shall endeavour to harmonise their national policies in this regard and for this purpose appoint the National Authority;

(d) The Contracting States shall co-operate with the competent international, regional and subregional organizations to establish and adopt regional standards, recommended practices and procedures to prevent, abate and combat pollution from all sources in conformity with the objectives of the present Convention, and to assist each other in fulfilling their obligations under the present Convention;

(e) The Contracting Series shall use their best endeavour to ensure that the implementation of the present Convention shall not cause transformation of one type of pollution to another which could be more detrimental to the environment.

Article IV: Pollution from Ships

The Contracting States shall take all appropriate measures in conformity with the present Convention and the applicable rules of international law to prevent, abate and combat pollution in the Sea Area caused by intentional or accidental discharges from ships, and shall ensure effective compliance in the Sea Area with applicable international rules relating to the control of this type of pollution, including load-on-top, segregated ballast and crude oil washing procedures for tankers.

Article V: Pollution Caused by Dumping from Ships and Aircraft

The Contracting States shall take all appropriate measures to prevent, abate and combat pollution in the Sea Area caused by dumping of wastes and other matter from ships and aircraft, and shall ensure effective compliance in the Sea Area with applicable international rules relating to the control of this type of pollution as provided for in relevant international conventions.

Article VI: Pollution from Land-Based Sources

The Contracting States shall take all appropriate measures to prevent, abate and combat pollution caused by discharges from land reaching the Sea Area whether water-borne, air-borne, or directly from the coast including outfalls and pipelines.

Article VII: Pollution Resulting from Exploration and Exploitation of the Bed of the Territorial Sea and Its Sub-Soil and the Continental Shelf

The Contracting States shall take all appropriate measures to prevent, abate and combat pollution in the Sea Area resulting from exploration and exploitation of the bed of the territorial sea and its sub-soil and the continental shelf, including the prevention of accidents and the combating of pollution emergencies resulting in damage to the marine environment.

Article VIII: Pollution from Other Human Activities

The Contracting States shall take all appropriate measures to prevent, abate and combat pollution of the Sea Area resulting from land reclamation and associated suction dredging and coastal dredging.

Article IX: Co-operation in Dealing With Pollution Emergencies

(a) The Contracting States shall, individually and/or jointly, take all necessary measures, including those to ensure that adequate equipment and qualified personnel are readily available, to deal with pollution emergencies in the Sea Area, whatever the cause of such emergencies, and to reduce or eliminate damage resulting therefrom;

(b) Any Contracting State which becomes aware of any pollution emergency in the Sea Area shall, without delay, notify the Organization referred to under Article XVI and, through the secretariat any Contracting State likely to be affected by such emergency.

Source: *Kuwait Regional Convention for Co-operation on the Protection of the Marine Environment from Pollution*, Apr. 24, 1978, 1140 U.N.T.S. 133, 17 I.L.M. 511 (1978).

Law of the Sea Convention

Part Twelve of the United Nations Convention on the Law of the Sea (UNCLOS) is devoted to environmental protection provisions, most of which are general in scope. In addition, Article 145 addresses pollution caused by mining and mineral retrieval from the deep seabed. Deep seabed mining was a significant concern in the drafting of UNCLOS; Part Eleven of the convention is devoted to it. In practice, as sometimes happens, the precautionary effort put into drafting provisions to regulate it appears to have been wasted. Deep seabed mining as an industry has not materialized. Yet the International Seabed Authority, with three *dozen employees and an annual budget of over $4 million, continues to exist and to draw up regulations for a nonexistent industry.*

Article 1: Use of terms and scope

1. For the purposes of this Convention:

(1) "Area" means the seabed and ocean floor and subsoil thereof, beyond the limits of national jurisdiction;

(2) "Authority" means the International Seabed Authority;

(3) "Activities in the Area" means all activities of exploration for, and exploitation of, the resources of the Area;

***

Article 145: Protection of the marine environment

Necessary measures shall be taken in accordance with this Convention with respect to activities in the Area to ensure effective protection for the marine environment from harmful effects which may arise from such activities. To this end the Authority shall adopt appropriate rules, regulations and procedures for inter alia:

(a) The prevention, reduction and control of pollution and other hazards to the marine environment, including the coastline, and of interference with the ecological balance of the marine environment, particular attention being paid to the need for protection from harmful effects of such activities as drilling, dredging, excavation, disposal of waste, construction and operation or maintenance of installations, pipelines and other devices related to such activities;

(b) The protection and conservation of the natural resources of the Area and the prevention of damage to the flora and fauna of the marine environment.

***

Part XII: Protection and Preservation of the Marine Environment

Section 1. General Provisions

Article 192: General obligation

States have the obligation to protect and preserve the marine environment.

Article 193: Sovereign right of States to exploit their natural resources

States have the sovereign right to exploit their natural resources pursuant to their environmental policies and in accordance with their duty to protect and preserve the marine environment.

Article 194: Measures to prevent, reduce and control pollution of the marine environment

1. States shall take, individually or jointly as appropriate, all measures consistent with this Convention that are necessary to prevent, reduce and control pollution of the marine environment from any source, using for this purpose the best practicable means at their disposal and in accordance with their capabilities, and they shall endeavour to harmonize their policies in this connection.

2. States shall take all measures necessary to ensure that activities under their jurisdiction or control are so conducted as not to cause damage by pollution to other States and their environment, and that pollution arising from incidents or activities under their jurisdiction or control does not spread beyond the areas where they exercise sovereign rights in accordance with this Convention.

3. The measures taken pursuant to this Part shall deal with all sources of pollution of the marine environment. These measures shall include, inter alia, those designed to minimize to the fullest possible extent:

(a) the release of toxic, harmful or noxious substances, especially those which are persistent, from land-based sources, from or through the atmosphere or by dumping;

(b) pollution from vessels, in particular measures for preventing accidents and dealing with emergencies, ensuring the safety of operations at sea, preventing intentional and unintentional discharges, and regulating the design, construction, equipment, operation and manning of vessels;

(c) pollution from installations and devices used in exploration or exploitation of the natural resources of the seabed and subsoil, in particular measures for preventing accidents and dealing with emergencies, ensuring the safety of operations at sea, and regulating the design, construction, equipment, operation and manning of such installations or devices;

(d) Pollution from other installations and devices operating in the marine environment, in particular measures for preventing accidents and dealing with emergencies, ensuring the safety of operations at sea, and regulating the design, construction, equipment, operation and manning of such installations or devices.

4. In taking measures to prevent, reduce or control pollution of the marine environment, States shall refrain from unjustifiable interference with activities carried out by other States in the exercise of their rights and in pursuance of their duties in conformity with this Convention.

5. The measures taken in accordance with this Part shall include those necessary to protect and preserve rare or fragile ecosystems as well as the habitat of depleted, threatened or endangered species and other forms of marine life.

Article 195: Duty not to transfer damage or hazards or transform one type of pollution into another

In taking measures to prevent, reduce and control pollution of the marine environment, States shall act so as not to transfer, directly or indirectly, damage or hazards from one area to another or transform one type of pollution into another.

Article 196: Use of technologies or introduction of alien or new species

1. States shall take all measures necessary to prevent, reduce and control pollution of the marine environment resulting from the use of technologies under their jurisdiction or control, or the intentional or accidental introduction of species, alien or new, to a particular part of the marine environment, which may cause significant and harmful changes thereto.

2. This article does not affect the application of this Convention regarding the prevention, reduction and control of pollution of the marine environment.

Section 2. Global and Regional Co-operation [omitted]

Section 3. Technical Assistance [omitted]

Section 4. Monitoring and Environmental Assessment [omitted]

Section 5. International Rules and National Legislation to Prevent, Reduce and Control Pollution of the Marine Environment

Article 207: Pollution from land-based sources

1. States shall adopt laws and regulations to prevent, reduce and control pollution of the marine environment from land-

based sources, including rivers, estuaries, pipelines and outfall structures, taking into account internationally agreed rules, standards and recommended practices and procedures.

2. States shall take other measures as may be necessary to prevent, reduce and control such pollution.

3. States shall endeavour to harmonize their policies in this connection at the appropriate regional level.

4. States, acting especially through competent international organizations or diplomatic conference, shall endeavour to establish global and regional rules, standards and recommended practices and procedures to prevent, reduce and control pollution of the marine environment from land-based sources, taking into account characteristic regional features, the economic capacity of developing States and their need for economic development. Such rules, standards and recommended practices and procedures shall be re-examined from time to time as necessary.

5. Laws, regulations, measures, rules, standards and recommended practices and procedures referred to in paragraphs 1, 2 and 4 shall include those designed to minimize, to the fullest extent possible, the release of toxic, harmful or noxious substances, especially those which are persistent, into the marine environment.

Article 208: Pollution from seabed activities subject to national jurisdiction

1. Coastal States shall adopt laws and regulations to prevent, reduce and control pollution of the marine environment arising from or in connection with seabed activities subject to their jurisdiction and from artificial islands, installations and structures under their jurisdiction, pursuant to articles 60 and 80.

2. States shall take other measures as may be necessary to prevent, reduce and control such pollution.

3. Such laws, regulations and measures shall be no less effective than international rules, standards and recommended practices and procedures.

4. States shall endeavour to harmonize their policies in this connection at the appropriate regional level.

5. States, acting especially through competent international organizations or diplomatic conference, shall establish global and regional rules, standards and recommended practices and procedures to prevent, reduce and control pollution of the marine environment referred to in paragraph 1. Such rules, standards and rec-

ommended practices and procedures shall be re-examined from time to time as necessary.

Article 209: Pollution from activities in the Area

1. International rules, regulations and procedures shall be established in accordance with Part XI to prevent, reduce and control pollution of the marine environment from activities in the Area. Such rules, regulations and procedures shall be re-examined from time to time as necessary.

2. Subject to the relevant provisions of this section, States shall adopt laws and regulations to prevent, reduce and control pollution of the marine environment from activities in the Area undertaken by vessels, installations, structures and other devices flying their flag or of their registry or operating under their authority, as the case may be. The requirements of such laws and regulations shall be no less effective than the international rules, regulations and procedures referred to in paragraph 1.

Article 210: Pollution by dumping

1. States shall adopt laws and regulations to prevent, reduce and control pollution of the marine environment by dumping.

2. States shall take other measures as may be necessary to prevent, reduce and control such pollution.

3–6. [omitted]

Article 211: Pollution from vessels

1. States, acting through the competent international organization or general diplomatic conference, shall establish international rules and standards to prevent, reduce and control pollution of the marine environment from vessels and promote the adoption, in the same manner, wherever appropriate, of routing systems designed to minimize the threat of accidents which might cause pollution of the marine environment, including the coastline, and pollution damage to the related interests of coastal States. Such rules and standards shall, in the same manner, be re-examined from time to time as necessary.

2. States shall adopt laws and regulations for the prevention, reduction and control of pollution of the marine environment from vessels flying their flag or of their registry. Such laws and regulations shall at least have the same effect as that of gen-

erally accepted international rules and standards established through the competent international organization or general diplomatic conference.

3–7. [omitted]

[Articles 212–237, also regarding environmental protection, omitted]

Source: *United Nations Convention on the Law of the Sea*, Dec. 10, 1982, U.N. Doc. A/CONF.62/122, 21 I.L.M. 1261 (1982).

Outer Space Treaty

The Outer Space Treaty, with its guarantees of freedom of access to outer space for all nations, harks back to the freedom-of-navigation treaties of the colonial era. Only a few countries are in a position to enjoy the freedom of exploration and use of outer space guaranteed by the treaty. So far outer space has turned out to be commercially worthless, with one exception: orbits near the earth are valuable for communications and observation satellites. At one time, the geostationary orbits above the equator were perceived as particularly valuable; the countries below these orbits (and thus with an arguable territorial claim to them) were, for the most part, not capable of launching satellites, while the countries that were capable of reaching those orbits were not located beneath them.

Article I

The exploration and use of outer space, including the moon and other celestial bodies, shall be carried out for the benefit and in the interests of all countries, irrespective of their degree of economic or scientific development, and shall be the province of all mankind.

Outer space, including the moon and other celestial bodies, shall be free for exploration and use by all states without discrimination of any kind, on a basis of equality and in accordance with international law, and there shall be free access to all areas of celestial bodies.

There shall be freedom of scientific investigation in outer space, including the moon and other celestial bodies, and states shall facilitate and encourage international co-operation in such investigation.

Article II

Outer space, including the moon and other celestial bodies, is not subject to national appropriation by claim of sovereignty, by means of use or occupation, or by any other means.

Source: *Treaty on Principles Governing the Activities of States in the Exploration and Use of Outer Space, Including the Moon and Other Celestial Bodies,* Oct. 10, 1967, 610 U.N.T.S. 205, 6 I.L.M. 386 (1967).

Protocol I to the Geneva Conventions of 1949

ENMOD, excerpted above, was one of two global attempts during the 1970s to address ecological harm from warfare. The other was Protocol I. Unlike ENMOD, Protocol I does not deal entirely or even primarily with environmental harm. Articles 35 and 55, though, contain direct environmental protections, while other articles, such as Article 54, contain indirect protections.

Article 35. Basic rules

1. In any armed conflict, the right of the parties to the conflict to choose methods or means of warfare is not unlimited.

2. It is prohibited to employ weapons, projectiles and material and methods of warfare of a nature to cause superfluous injury or unnecessary suffering.

3. It is prohibited to employ methods or means of warfare which are intended, or may be expected, to cause widespread, long-term and severe damage to the natural environment.

***

Article 54. Protection of objects indispensable to the survival of the civilian population

1. Starvation of civilians as a method of warfare is prohibited.

2. It is prohibited to attack, destroy, remove or render useless objects indispensable to the survival of the civilian population, such as foodstuffs, agricultural areas for the production of foodstuffs, crops, livestock, drinking water installations and supplies and irrigation works, for the specific purpose of denying them for their sustenance value to the civilian population or to the adverse party, whatever the motive, whether in order to starve out civilians, to cause them to move away, or for any other motive.

3. The prohibitions in paragraph 2 shall not apply to such of the objects covered by it as are used by an adverse Party:

(a) As sustenance solely for the members of its armed forces; or

(b) If not as sustenance, then in direct support of military action, provided, however, that in no event shall actions against these objects be taken which may be expected to leave the civilian population with such inadequate food or water as to cause its starvation or force its movement.

4. These objects shall not be made the object of reprisals.

5. In recognition of the vital requirements of any party to the conflict in the defense of its national territory against invasion, derogation from the prohibitions contained in paragraph 2 may be made by a party to the conflict within such territory under its own control where required by imperative military necessity.

Article 55. Protection of the natural environment

1. Care shall be taken in warfare to protect the natural environment against widespread, long-term and severe damage. This protection includes a prohibition of the use of methods or means of warfare which are intended or may be expected to cause such damage to the natural environment and thereby to prejudice the health or survival of the population.

2. Attacks against the natural environment by way of reprisals are prohibited.

Source: *Protocol Additional to the Geneva Conventions of 12 August 1949, and Relating to the Protection of Victims of International Armed Conflicts*, June 8, 1977, 1125 U.N.T.S. 3.

Protocol II to the Geneva Conventions of 1949

A sad feature of modern warfare is that most armed conflicts since World War II have been internal. Traditional wars between sovereign states, the type of conflict addressed by the bulk of international law on the subject, have been less common than armed insurrections, revolutions, wars of national independence, coups d'etat, civil wars, ethnic strife, and other internal armed conflicts; yet the law governing those conflicts is far less developed, in part because traditionally international law governed relations between states rather than things happening within the borders of a single state.

The suffering of the civilian population is the same, whether the war in which they are unwillingly caught up is international or inter-

nal; the difference in protection comes not from any perception that the victims of internal conflicts are somehow less worthy of protection, but from the reluctance of states to allow the same degree of international scrutiny of and potential interference in affairs within their borders as they are willing to allow of their actions beyond those borders—and of potential actions by foreign sovereigns within them.

By 1949, the need for some sort of law to address internal armed conflicts was apparent, and Common Article 3 of the four Geneva conventions does so. The protection in Common Article 3 was expanded upon and added to in the 1970s with Protocol II. Neither Common Article 3 nor Protocol II contains any equivalent of Protocol I's Articles 35 and 55, however. Protocol II's Article 14, reproduced below in its entirety, contains protections similar to, but less extensive than, those in Protocol I's Article 54. Compare the two articles to see the difference in coverage.

Article 14. Protection of objects indispensable to the survival of the civilian population

Starvation of civilians as a method of combat is prohibited. It is therefore prohibited to attack, destroy, remove, or render useless, for that purpose, objects indispensable to the survival of the civilian population, such as foodstuffs, agricultural areas for the production of foodstuffs, crops, livestock, drinking water installations and supplies, and irrigation works.

Source: *Protocol Additional to the Geneva Conventions of 12 August 1949, and Relating to the Protection of Victims of Non-International Armed Conflicts,* June 8, 1977, 1125 U.N.T.S. 609.

Rome Statute on the International Criminal Court

The International Criminal Court has attracted a great deal of media attention in the United States because the United States has not become a party to the court's statute. Ninety-four countries have become parties to the Rome Statute; however, a majority of the world's population lives in countries that are not parties, including the United States and other large countries such as China, India, and Russia. Even for those nonparties, however, the list of war crimes contained in Article 8 seems likely to be considered definitive and comprehensive. These war crimes include environmental crimes; the language of Article 8(2)(b)(iv) incorporates Protocol I's "widespread, long-term, and severe damage" standard rather than ENMOD's "widespread, long-term or severe damage," and adds the qualification that even "widespread, long-term, and severe damage" is only a war crime if it "would be clearly excessive in relation to the concrete and direct overall military advantage anticipated and the attack was intentionally launched with the knowledge that such damage would result." The Rome Statute's definition of an environmental war crime is thus narrower than that in either ENMOD or Protocol I.

Article 8 also incorporates other quasi-environmental protections analogous or identical to those in the Geneva Conventions and Protocol I and the Hague Conventions. As with Common Article 3 of the Geneva Conventions and Protocol II, any protection given to the environment and victims of environmental damage in internal armed conflicts is incidental.

Article 8: War crimes

1. The Court shall have jurisdiction in respect of war crimes in particular when committed as part of a plan or policy or as part of a large-scale commission of such crimes.

2. For the purpose of this Statute, "war crimes" means:

(a) Grave breaches of the Geneva Conventions of 12 August 1949.

(b) Other serious violations of the laws and customs applicable in international armed conflict, within the established framework of international law, namely, any of the following acts:

***

(iv) Intentionally launching an attack in the knowledge that such attack will cause incidental loss of life or injury to civilians or damage to civilian objects or widespread, long-term and severe damage to the natural environment which would be clearly excessive in relation to the concrete and direct overall military advantage anticipated;

***

(xvii) Employing poison or poisoned weapons;

(xviii) Employing asphyxiating, poisonous or other gases, and all analogous liquids, materials or devices;

***

(xxv) Intentionally using starvation of civilians as a method of warfare by depriving them of objects indispensable to their survival, including willfully impeding relief supplies as provided for under the Geneva Conventions;

***

(c) In the case of an armed conflict not of an international character, serious violations of article 3 common to the four Geneva Conventions of 12 August 1949, namely, any of the following acts committed against persons taking no active part in the hostilities, including members of armed forces who have laid down their arms and those placed hors de combat by sickness, wounds, detention or any other cause:

(i) Violence to life and person, in particular murder of all kinds, mutilation, cruel treatment and torture;

(ii) Committing outrages upon personal dignity, in particular humiliating and degrading treatment;

(iii) Taking of hostages;

(iv) The passing of sentences and the carrying out of executions without previous judgement pronounced by a regularly constituted court, affording all judicial guarantees which are generally recognized as indispensable.

Source: *Rome Statute on the International Criminal Court*, U.N. Doc. A/ CONF. 183/9 (1998).

International Courts and Tribunals

Corfu Channel Case

The time immediately after the end of World War II was a time of great international tension in Europe, as the lines of the Cold War were being drawn and the possibility of another eruption of international conflict remained present. Against this background, in late 1946 two British warships struck mines in the channel dividing the Greek island of Corfu from Albania on the European mainland. The mines had apparently been placed there with the knowledge of the Albanian government, which had then failed to warn other countries of the minefield and the danger it presented to shipping in the Corfu Channel. In addressing the question of Albania's responsibility for the harm to the United Kingdom, the International Court of Justice stated the basic principle of state responsibility in what has become the most frequently quoted phrase in the opinion: it is "every State's obligation not to allow knowingly its territory to be used for acts contrary to the rights of other States."

From all the facts and observations mentioned above, the Court draws the conclusion that the laying of the minefield which caused the explosions on October 22nd, 1946, could not have been accomplished without the knowledge of the Albanian Government.

The obligations resulting for Albania from this knowledge are not disputed between the Parties. Counsel for the Albanian Government expressly recognized that [*translation*] "if Albania had been informed of the operation before the incidents of October 22nd, and in time to warn the British vessels and shipping in general of the existence of mines in the Corfu Channel, her responsibility would be involved. . . ."

The obligations incumbent upon the Albanian authorities consisted in notifying, for the benefit of shipping in general, the existence of a minefield in Albanian territorial waters and in warning the approaching British warships of the imminent danger to which the minefield exposed them. Such obligations are based, not on the Hague Convention of 1907, No. VIII, which is applicable in time of war, but on certain general and well-recognized principles, namely: elementary considerations of humanity, even more exacting in peace than in war; the principle of the freedom of maritime communication; and every State's obligation not to allow knowingly its territory to be used for acts contrary to the rights of other States.

In fact, Albania neither notified the existence of the minefield, nor warned the British warships of the danger they were approaching.

Source: *Corfu Channel Case (U.K. v. Alb.)*, 1949 I.C.J. 4, 22 (1949) (determination on the merits).

Gabčíkovo-Nagymaros Case

The Gabčíkovo-Nagymaros case, decided after the Rio Declaration's comprehensive statement of the principles of sustainable development had been widely disseminated, gave the International Court of Justice (ICJ) an opportunity to address whether those principles had, in fact, attained the status of customary international law in the same manner as Principle 21 of the Stockholm Declaration. The ICJ decided that they had not. In a separate opinion, however, the vice president of the court, Sri Lankan judge Christopher Gregory Weeramantry, maintained that in fact sustainable development had acquired the status of customary international law. The principles of sustainable development had, according to Judge Weeramantry, either expressed pre-existing norms or had given rise to subsequent normative expectations about the behavior of states and were widely accepted in practice. In reaching this conclusion, he relied not only on the Rio Declaration and other relatively recent statements of sustainable development principles, but also on historical examples from several different cultural regions of the world stretching over thousands of years. The historical portion of the opinion, although fascinating, is too lengthy to reproduce here. Judge Weeramantry's opinion may, however, be read in its entirety on the website of the International Court of Justice at http://www.icj-cij.org/icjwww/icases/iunan/iunan_judgment_advisory%20opinion_19960708/iunan_ijudgment_advisory%20opinion_19960708_Opinions/iunan_ijudgment_19 960708_Dissenting_Weeramantry.htm (visited September 8, 2004).

[From the Opinion of the Court]:

140. It is clear that the Project's impact upon, and its implications for, the environment are of necessity a key issue. The numerous scientific reports which have been presented to the Court by the Parties—even if their conclusions are often contradictory—provide abundant evidence that this impact and these implications are considerable.

***

Throughout the ages, mankind has, for economic and other reasons, constantly interfered with nature. In the past, this was often done without consideration of the effects upon the environment. Owing to new scientific insights and to a growing awareness of the risks for mankind—for present and future generations—of pursuit of such interventions at an unconsidered and unabated pace, new norms and standards have been developed, set forth in a great number of instruments during the last two decades. Such new norms have to be taken into consideration, and such new standards given proper weight, not only when States contemplate new activities but also when continuing with activities begun in the past. This need to reconcile economic development with protection of the environment is aptly expressed in the concept of sustainable development.

***

Separate Opinion of Vice-President Weeramantry

***

The Concept of Sustainable Development

***

The Court has referred to [sustainable development] as a concept in paragraph 140 of its Judgment. However, I consider it to be more than a mere concept, but as a principle with normative value which is crucial to the determination of this case. Without the benefits of its insights, the issues involved in this case would have been difficult to resolve.

Since sustainable development is a principle fundamental to the determination of the competing considerations in this case, and since, although it has attracted attention only recently in the literature of international law, it is likely to play a major role in determining important environmental disputes of the future, it calls for consideration in some detail. Moreover, this is the first occasion on which it has received attention in the jurisprudence of this Court.

***

The present case . . . focuses attention, as no other case has done in the jurisprudence of this Court, on the question of the harmonization of developmental and environmental concepts.

(a) *Development as a Principle of International Law.* Article 1 of the Declaration on the Right to Development, 1986, asserted that "The right to development is an inalienable human right." This Declaration had the overwhelming support of the international community and has been gathering strength since then. Principle 3 of the Rio Declaration, 1992, reaffirmed the need for the right to development to be fulfilled.

"Development" means, of course, development not merely for the sake of development and the economic gain it produces, but for its value in increasing the sum total of human happiness and welfare. That could perhaps be called the first principle of the law relating to development.

To the end of improving the sum total of human happiness and welfare, it is important and inevitable that development projects of various descriptions, both minor and major, will be launched from time to time in all parts of the world.

(b) *Environmental Protection as a Principle of International Law.* The protection of the environment is likewise a vital part of contemporary human rights doctrine, for it is a sine qua non for numerous human rights such as the right to health and the right to life itself. It is scarcely necessary to elaborate on this, as damage to the environment can impair and undermine all the human rights spoken of in the Universal Declaration and other human rights instruments.

While, therefore, all peoples have the right to initiate development projects and enjoy their benefits, there is likewise a duty to ensure that those projects do not significantly damage the environment.

(c) *Sustainable Development as a Principle of International Law.* After the early formulations of the concept of development, it has been recognized that development cannot be pursued to such a point as to result in substantial damage to the environment within which it is to occur. Therefore development can only be prosecuted in harmony with the reasonable demands of environmental protection. Whether development is sustainable by reason of its impact on the environment will, of course, be a question to be answered in the context of the particular situation involved.

It is thus the correct formulation of the right to development that that right does not exist in the absolute sense, but is relative always to its tolerance by the environment. The right to development as thus refined is clearly part of modern international law. It is compendiously referred to as sustainable development.

The concept of sustainable development can be traced back, beyond the Stockholm Conference of 1972, to such events as the Founex meeting of experts in Switzerland in June 1971; the conference on environment and development in Canberra in 1971; and United Nations General Assembly Resolution 2849 (XXVI). It received a powerful impetus from the Stockholm Declaration which, by Principle 11, stressed the essentiality of development as well as the essentiality of bearing environmental considerations in mind in the developmental process. Moreover, many other Principles of that Declaration provided a setting for the development of the concept of sustainable development and more than one-third of the Stockholm Declaration related to the harmonization of environment and development. The Stockholm Conference also produced an Action Plan for the Human Environment.

The international community had thus been sensitized to this issue even as early as the early 1970s . . . .

***

. . . In 1992, the Rio Conference made it a central feature of its Declaration, and it has been a focus of attention in all questions relating to development in the developing countries.

The principle of sustainable development is thus a part of modern international law by reason not only of its inescapable logical necessity, but also by reason of its wide and general acceptance by the global community.

***

Evidence appearing in international instruments and State practice (as in development assistance and the practice of international financial institutions) likewise amply supports a contemporary general acceptance of the concept.

Recognition of the concept could thus, fairly, be said to be worldwide.

(d) The Need for International Law to Draw upon the World's Diversity of Cultures in Harmonizing Development and Environmental Protection.

***

. . . Environmental law is now in a formative stage, not unlike international law in its early stages. A wealth of past experience from a variety of cultures is available to it. It would be pity indeed if it were left untapped merely because of attitudes of formalism which see such approaches as not being entirely *de rigueur.*

***

Especially where this Court is concerned, "the essence of true universality" of the institution is captured in the language of Article 9 of the Statute of the International Court of Justice which requires the "representation of the *main forms of civilization* and of the principal legal systems of the world" (emphasis added). The struggle for the insertion of the italicized words in the Court's Statute was a hard one, led by the Japanese representative, Mr. Adatci, and, since this concept has thus been integrated into the structure and the Statute of the Court, I see the Court as being charged with a duty to draw upon the wisdom of the world's several civilizations, where such a course can enrich its insights into the matter before it. The Court cannot afford to be monocultural, especially where it is entering newly developing areas of law.

[Vice President Weeramantry then discusses historical examples of sustainable development, and their roots in traditional philosophies, from

Sri Lanka, the Sonjo and Chagga cultures of Tanzania, Iran, China, and the Inca civilization of South America, with passing references to almost all of the world. The discussion is interesting but too lengthy to reproduce here.]

Source: *Gabčíkovo-Nagymaros Dispute (Slovakia v. Hungary)*, 1997 I.C.J. 7 (1997).

Trail Smelter Arbitration

The second Trail Smelter Arbitral Award is perhaps the best-known international environmental law decision to date. While in hindsight the arbitral tribunal's statement, in dicta, of the principle in the excerpt below may seem obvious and unadventurous, it marks the beginning of modern international environmental law.

The Trail Smelter case is discussed in Chapter 1. The facts are fairly simple, although the details are complex: A smelter in Trail, British Columbia, released fumes that caused environmental damage to territory in the state of Washington, on the other side of the international border. The United States sought an end to the transboundary pollution, requesting that Canada be required to refrain from allowing the Trail smelter to operate in a way that would cause further damage to the territory of the United States.

The Trail smelter—the actual smelting plant, that is—is still operating today. The facilities have been upgraded and rebuilt many times, but the smelter, now over a century old, continues to be a focus of concern for environmental activists. In 1997, as the Trail smelter entered its second century, the company that now owns it, Teck Cominco, completed a twenty-year, billion-dollar series of upgrades by installing a new smelter at the site. The new smelter operates far more cleanly, but a century of pollution has left the soil, water, and wildlife downstream and downwind from the smelter heavily contaminated with lead and other toxic substances. The pollution is not confined to the Canadian side of the border; parts of Washington state are contaminated as well.

The Trail smelter also illustrates the tension between environment and development: Teck Cominco's operations at Trail, centered on the Trail smelter, are the largest employer in the Trail area. As late as 2001, children in Trail had levels of lead in their blood high enough to reduce intelligence, stunt growth, and cause neurological disorders. Additional environmental improvements in 2001 have reduced this problem.

[From the Award of March 11, 1941]:

***

The second question under Article III of the Convention is as follows:

In the event of the answer to the first part of the preceding question being in the affirmative, whether the Trail Smelter should be required to refrain from causing damage in the State of Washington in the future and, if so, to what extent?

Damage has occurred since January 1, 1932, as fully set forth in the previous decision. To that extent, the first part of the preceding question has thus been answered in the affirmative.

As has been said above, the report of the International Joint Commission (1 (g)) contained a definition of the word "damage" excluding "occasional damage that may be caused by SO_2 fumes being carried across the international boundary in air pockets or by reason of unusual atmospheric conditions," as far, at least, as the duty of the Smelter to reduce the presence of that gas in the air was concerned.

***

Great progress in the control of fumes has been made by science in the last few years and this progress should be taken into account.

The Tribunal, therefore, finds . . . that, under the principles of international law, as well as of the law of the United States, no State has the right to use or permit the use of its territory in such a manner as to cause injury by fumes in or to the territory of another or the properties or persons therein, when the case is of serious consequence and the injury is established by clear and convincing evidence.

Source: *Trail Smelter Case (U.S. v. Can.)*, 3 R.I.A.A. 1905, 1965 (1941), reprinted in 35 *Am. Journal of International Law* 684 (1941).

Final Report of the Office of the Prosecutor for the International Criminal Court for the Former Yugoslavia Regarding Possible NATO War Crimes

The 1999 war between the North Atlantic Treaty Organization (NATO) and Yugoslavia was subjected to an unprecedented level of legal scrutiny as it actually took place. The parties to the conflict, as well as various international bodies, considered a wide variety of legal claims,

*including the possibility that some of NATO's acts constituted environ-
mental war crimes. In the case it brought before the International Court
of Justice, Yugoslavia (now Serbia and Montenegro) included environmental crimes among the harms it alleged to have been committed by the
NATO parties. Cases against two of those parties (Spain and the United
States) were promptly dismissed for lack of jurisdiction, but cases
against eight more (Belgium, Canada, France, Germany, Italy, the
Netherlands, Portugal, and the United Kingdom) remain pending.*

*The ICJ has jurisdiction over actions between states; another inter-
national court, the International Criminal Tribunal for the Former Yugoslavia (ICTY) has jurisdiction over war crimes committed by individuals. Yugoslavia, as well as activists in the United States and Europe,
urged the ICTY's Office of the Prosecutor to investigate the possibility
that some of NATO's acts constituted war crimes, including environmental war crimes.*

*The excerpt below is from the report of the Office of the Prosecutor;
it is not a decision of an international court. The ICTY made no decision
on NATO's war crimes, because no prosecution for those crimes was
ever brought before it. The excerpt below addresses environmental war
crimes; as can be seen, the Office of the Prosecutor recommended that
the then-prosecutor, Carla Del Ponte, bring no case.*

IV. Assessment: General Issues
Damage to the Environment

14. The NATO bombing campaign did cause some damage
to the environment. For instance, attacks on industrial facilities
such as chemical plants and oil installations were reported to
have caused the release of pollutants, although the exact extent
of this is presently unknown. The basic legal provisions applicable to protection of the environment in armed conflict are Article 35(3) of Additional Protocol I, which states that '[i]t is prohibited to employ methods or means of warfare which are
intended, or may be expected, to cause widespread, long-term
and severe damage to the natural environment' and Article 55
which states:

(1) Care shall be taken in warfare to protect the natural
environment against widespread, long-term, and severe damage.
This protection includes a prohibition of the use of methods or

means of warfare which are intended or may be expected to cause such damage to the natural environment and thereby to prejudice the health or survival of the population.

(2) Attacks against the natural environment by way of reprisals are prohibited.

15. Neither the USA nor France has ratified Additional Protocol I. Article 55 may, nevertheless, reflect current customary law (see however the 1996 Advisory Opinion on the Legality of Nuclear Weapons, where the International Court of Justice appeared to suggest that it does not (ICJ Rep. (1996), 242, para. 31)). In any case, Articles 35(3) and 55 have a very high threshold of application. Their conditions for application are extremely stringent and their scope and contents imprecise. For instance, it is generally assumed that Articles 35(3) and 55 only cover very significant damage. The adjectives "widespread, long-term, and severe" used in Additional Protocol I are joined by the word "and," meaning that it is a triple, cumulative standard that needs to be fulfilled.

Consequently, it would appear extremely difficult to develop a *prima facie* case upon the basis of these provisions, even assuming they were applicable. For instance, it is thought that the notion of 'long-term' damage in Additional Protocol I would need to be measured in years rather than months, and that, as such, ordinary battlefield damage of the kind caused to France in World War I would not be covered.

The great difficulty of assessing whether environmental damage exceeded the threshold of Additional Protocol I has also led to criticism by ecologists. This may partly explain the disagreement as to whether any of the damage caused by the oil spills and fires in the 1990/91 Gulf War technically crossed the threshold of Additional Protocol I.

It is the committee's view that similar difficulties would exist in applying Additional Protocol I to the present facts, even if reliable environmental assessments were to give rise to legitimate concern concerning the impact of the NATO bombing campaign. Accordingly, these effects are best considered from the underlying principles of the law of armed conflict such as necessity and proportionality.

16. The conclusions of the Balkan Task Force (BTF) established by UNEP to look into the Kosovo situation are:

"Our findings indicate that the Kosovo conflict has not caused an environmental catastrophe affecting the Balkans region as a whole.

Nevertheless, pollution detected at some sites is serious and poses a threat to human health.

BTF was able to identify environmental 'hot spots,' namely in Pančevo, Kragujevac, Novi Sad and Bor, where immediate action and also further monitoring and analyses will be necessary. At all of these sites, environmental contamination due to the consequences of the Kosovo conflict was identified.

Part of the contamination identified at some sites clearly predates the Kosovo conflict, and there is evidence of long-term deficiencies in the treatment and storage of hazardous waste.

The problems identified require immediate attention, irrespective of their cause, if further damage to human health and the environment is to be avoided."

17. The OTP has been hampered in its assessment of the extent of environmental damage in Kosovo by a lack of alternative and corroborated sources regarding the extent of environmental contamination caused by the NATO bombing campaign. Moreover, it is quite possible that, as this campaign occurred only a year ago, the UNEP study may not be a reliable indicator of the long term environmental consequences of the NATO bombing, as accurate assessments regarding the long-term effects of this contamination may not yet be practicable.

It is the opinion of the committee, on the basis of information currently in its possession, that the environmental damage caused during the NATO bombing campaign does not reach the Additional Protocol I threshold. In addition, the UNEP Report also suggests that much of the environmental contamination which is discernible cannot unambiguously be attributed to the NATO bombing.

18. The alleged environmental effects of the NATO bombing campaign flow in many cases from NATO's striking of legitimate military targets compatible with Article 52 of Additional Protocol I such as stores of fuel, industries of fundamental importance for the conduct of war and for the manufacture of supplies and material of a military character, factories or plant and manufacturing centres of fundamental importance for the conduct of war. Even when targeting admittedly legitimate military objectives, there is

a need to avoid excessive long-term damage to the economic infrastructure and natural environment with a consequential adverse effect on the civilian population. Indeed, military objectives should not be targeted if the attack is likely to cause collateral environmental damage which would be excessive in relation to the direct military advantage which the attack is expected to produce (A.P.V. Rogers, "Zero Casualty Warfare," IRRC, March 2000, Vol. 82, pp. 177–178).

***

25. It is therefore the opinion of the committee, based on information currently available to it, that the OTP should not commence an investigation into the collateral environmental damage caused by the NATO bombing campaign.

Use of Depleted Uranium Projectiles

26. There is evidence of use of depleted uranium (DU) projectiles by NATO aircraft during the bombing campaign. There is no specific treaty ban on the use of DU projectiles. There is a developing scientific debate and concern expressed regarding the impact of the use of such projectiles and it is possible that, in future, there will be a consensus view in international legal circles that use of such projectiles violates general principles of the law applicable to use of weapons in armed conflict. No such consensus exists at present. Indeed, even in the case of nuclear warheads and other weapons of mass-destruction—those which are universally acknowledged to have the most deleterious environmental consequences—it is difficult to argue that the prohibition of their use is in all cases absolute [*Legality of Nuclear Weapons,* ICJ Rep. 242 (1996)]. In view of the uncertain state of development of the legal standards governing this area, it should be emphasised that the use of depleted uranium or other potentially hazardous substance by any adversary to conflicts within the former Yugoslavia since 1991 has not formed the basis of any charge laid by the Prosecutor. It is acknowledged that the underlying principles of the law of armed conflict such as proportionality are applicable also in this context; however, it is the committee's view that analysis undertaken above (paras. 14–25) with regard to environmental damage would apply, mutatis mutandis, to the use of depleted uranium projectiles by NATO. It is therefore the opinion of

the committee, based on information available at present, that the OTP should not commence an investigation into use of depleted uranium projectiles by NATO.

Source: *Final Report to the Prosecutor by the Committee Established to Review the NATO Bombing Campaign Against the Federal Republic of Yugoslavia*, at http://www.un.org/icty/pressreal/nato061300.htm (visited May 28, 2004).

Other International Materials
Bogotá Declaration

The Outer Space Treaty, excerpted earlier in this chapter, guaranteed freedom of exploration and use of outer space to all nations. Yet few things in outer space seemed to have any actual value. One exception was geostationary orbit positions. A satellite that orbits the Earth at an altitude of 22,300 miles above the equator will circle the planet's center at the same rate as a point on the equator does: once per day. From the point of view of an observer on the Earth's surface, the satellite will appear to be motionless, forever fixed directly above the same point.

This motionlessness relative to the Earth's surface has advantages for communications satellites; it is far easier to transmit signals to and receive signals from a satellite if its location does not change. Yet under the Outer Space Treaty, those orbits are not subject to the sovereignty of the country below them. While from the perspective of nonequatorial countries, which are a majority of the world's countries and include the overwhelming majority of the world's people, this may seem perfectly fair, the perception of the equatorial countries was that a valuable resource was being taken from them, without compensation, and made available to the developed countries—neocolonialism in outer space. The Bogotá Declaration, signed by Brazil, Colombia, Congo (Republic of the Congo), Ecuador, Indonesia, Kenya, Uganda, and Zaire (now Democratic Republic of the Congo), attempted to address this perceived injustice.

The undersigned representatives of the States traversed by the Equator met in Bogotá, Republic of Colombia, from 29 November through 3 December, 1976 with the purpose of studying the geostationary orbit that corresponds to their national terrestrial, sea, and insular territory and considered as a natural resource. After an exchange of information and having studied in detail the different technical, legal, and political aspects implied in the exercise of national sovereignty of States adjacent to the said orbit, have reached the following conclusions:

1. The Geostationary Orbit as a Natural Resource

The geostationary orbit is a circular orbit on the Equatorial plane in which the period of sidereal revolution of the satellite is equal to the period of sidereal rotation of the Earth and the satellite moves in the same direction of the Earth's rotation. When a satellite describes this particular orbit, it is said to be geostationary; such a satellite appears to be stationary in the sky, when viewed from the earth, and is fixed on the zenith of a given point of the Equator, whose longitude is by definition that of the satellite.

This orbit is located at an approximate distance of 35,871 kilometers over the Earth's Equator.

Equatorial countries declare that the geostationary synchronous orbit is a physical fact linked to the reality of our planet because its existence depends exclusively on its relation to gravitational phenomena generated by the Earth, and that is why it must not be considered part of the outer space. Therefore, the segments of geostationary synchronous orbit are part of the territory over which Equatorial states exercise their national sovereignty. The geostationary orbit is a scarce natural resource, whose importance and value increase rapidly together with the development of space technology and with the growing need for communication; therefore, the Equatorial countries meeting in Bogota have decided to proclaim and defend on behalf of their peoples, the existence of their sovereignty over this natural resource. The geostationary orbit represents a unique facility that it alone can offer for telecommunication services and other uses which require geostationary satellites.

The frequencies and orbit of geostationary satellites are limited natural resources, fully accepted as such by current standards of the International Telecommunications Union. Technological advancement has caused a continuous increase in the number of satellites that use this orbit, which could result in a saturation in the near future.

The solutions proposed by the International Telecommunications Union and the relevant documents that attempt to achieve a better use of the geostationary orbit that shall prevent its imminent saturation, are at present impracticable and unfair and would considerably increase the exploitation costs of this resource especially for developing countries that do not have equal technological and financial resources as compared to industrialized countries, who enjoy an apparent monopoly in the exploitation and use of its geostationary synchronous orbit. In spite of the principle established by Article 33, sub-paragraph 2 of the International Telecommunications Convention, of 1973, that in the use of frequency bands for space radio communications, the members shall take into account that the frequencies and the orbit for geostationary satellites are limited natural resources that must be used efficiently and economically to allow the equitable access to this orbit and to its frequencies, we can see that both the geostationary orbit and the frequencies have been used in a way that does not allow the equitable access of the developing countries that do not have the technical and financial means that the great powers have. Therefore, it is imperative for the equatorial countries to exercise their sovereignty over the corresponding segments of the geostationary orbit.

2. Sovereignty of Equatorial States over the Corresponding Segments of the Geostationary Orbit

In qualifying this orbit as a natural resource, equatorial states reaffirm "the right of the peoples and of nations to permanent sovereignty over their wealth and natural resources that must be exercised in the interest of their national development and of the welfare of the people of the nation concerned," as it is set forth in Resolution 2692 (XXV) of the United Nations General Assembly entitled "permanent sovereignty over the natural resources of developing countries and expansion of internal accumulation sources for economic developments."

Furthermore, the charter on economic rights and duties of states solemnly adopted by the United Nations General Assembly through Resolution 3281 (XXIV), once more confirms the existence of a sovereign right of nations over their natural resources, in Article 2 subparagraph i, which reads:

"All states have and freely exercise full and permanent sovereignty, including possession, use and disposal of all their wealth, natural resources and economic activities."

Consequently, the above-mentioned provisions lead the equatorial states to affirm that the synchronous geostationary orbit, being a natural resource, is under the sovereignty of the equatorial states.

3. Legal State of the Geostationary Orbit

Bearing in mind the existence of sovereign rights over segments of geostationary orbit, the equatorial countries consider that the applicable legal consultations in this area must take into account the following:

(a) The sovereign rights put forward by the equatorial countries are directed towards rendering tangible benefits to their respective people and for the universal community, which is completely different from the present reality when the orbit is used to the greater benefit of the most developed countries.

(b) The segments of the orbit corresponding to the open sea are beyond the national jurisdiction of states will be considered as common heritage of mankind. Consequently, the competent international agencies should regulate its use and exploitation for the benefit of mankind.

(c) The equatorial states do not object to the free orbital transit of satellites approved and authorized by the International Telecommunications Convention, when these satellites pass through their outer space in their gravitational flight outside their geostationary orbit.

(d) The devices to be placed permanently on the segment of a geostationary orbit of an equatorial state shall require previous and expressed authorization on the part of the concerned state, and the operation of the device should conform with the national law of that territorial country over which it is placed. It must be understood that the said authorization is different from the co-ordination requested in cases of interference among satellite systems, which are specified in the regulations for radio communications. The said authorization refers in very clear terms to the countries' right to allow the operation of fixed radio communications stations within their territory.

(e) Equatorial states do not condone the existing satellites or the position they occupy on their segments of the Geostationary Orbit nor does the existence of said satellites confer any rights of placement of satellites or use of the segment unless expressly authorized by the state exercising sovereignty over this segment.

4. Treaty of 1967

The Treaty of 1967 on "The Principles Governing the Activities of States in the Exploration and Use of Outer Space, Including the Moon and Other Celestial Bodies," signed on 27 January 1967, cannot be considered as a final answer to the problem of the exploration and use of outer space, even less when the international community is questioning all the terms of international law which were elaborated when the developing countries could not count on adequate scientific advice and were thus not able to observe and evaluate the omissions, contradictions and consequences of the proposals which were prepared with great ability by the industrialized powers for their own benefit.

There is no valid or satisfactory definition of outer space which may be advanced to support the argument that the geostationary orbit is included in the outer space. The legal affairs subcommission which is dependent on the United Nations Commission on the Use of Outer Space for Peaceful Purposes, has been working for a long time on a definition of outer space, however, to date, there has been no agreement in this respect.

Therefore, it is imperative to elaborate a juridical definition of outer space, without which the implementation of the Treaty of 1967 is only a way to give recognition to the presence of the states that are already using the geostationary orbit. Under the name of a so-called non-national appropriation, what was actually developed was technological partition of the orbit, which is simply a national appropriation, and this must be denounced by the equatorial countries. The experiences observed up to the present and the development foreseeable for the coming years bring to light the obvious omissions of the Treaty of 1967 which force the equatorial states to claim the exclusion of the geostationary orbit.

The lack of definition of outer space in the Treaty of 1967, which has already been referred to, implies that Article II should not apply to geostationary orbit and therefore does not affect the right of the equatorial states that have already ratified the Treaty.

5. Diplomatic and Political Action

While Article 2 of the aforementioned Treaty does not establish an express exception regarding the synchronous geostationary orbit, as an integral element of the territory of equatorial states, the countries that have not ratified the Treaty should refrain from undertaking any procedure that allows the enforcement of provisions whose juridical omission has already been denounced.

The representatives of the equatorial countries attending the meeting in Bogotá, wish to clearly state their position regarding the declarations of Colombia and Ecuador in the United Nations, which affirm that they consider the geostationary orbit to be an integral part of their sovereign territory; this declaration is a historical background for the defense of the sovereign rights of the equatorial countries. These countries will endeavour to make similar declarations in international agencies dealing with the same subject and to align their international policy in accordance with the principles elaborated in this document.

Source: *Declaration of the First Meeting of the Equatorial Countries,* Dec. 3, 1976, I.T.U. Doc. WARC-BS 81-E, reprinted in Nicolas Matte, *Aerospace Law: Telecommunications Satellites,* 341–344 (1982); also available from http://www.jaxa.jp/jda/library/space-law/chapter_2/2-2-1-2_e.html (visited September 7, 2004)

Rio Declaration

Twenty years after the Stockholm Conference came the 1992 Earth Summit—the UN Conference on Environment and Development at Rio de Janeiro, Brazil. The two decades between the two conferences saw dramatic changes: advances in human understanding of environmental systems and processes, an increased ability on the part of the developing world to express its concerns, and the end of the Cold War that had dominated international relations in the post–World War II era.

The Rio Declaration on Environment and Development reflects these changes. Like the Stockholm Declaration, it sets out a series of Principles. Principle 1 clearly rejects a Deep Green or nature-first approach to sustainable development, opting instead for a humanocentric approach. Principle 2 restates Principle 21 of the Stockholm Declaration, with the addition of two words: "and development." The importance of development is again emphasized in several of the principles, including Principles 4, 5, 6, 12, and 25. Intergenerational equity, the precautionary principle, and the pollute-pays principle are set forth in Principles 3, 15, and 16, respectively.

Principle 1
Human beings are at the centre of concerns for sustainable development. They are entitled to a healthy and productive life in harmony with nature.

Principle 2
States have, in accordance with the Charter of the United Nations and the principles of international law, the sovereign right to exploit their own resources pursuant to their own environmental and developmental policies, and the responsibility to ensure that activities within their jurisdiction or control do not cause damage to the environment of other States or of areas beyond the limits of national jurisdiction.

Principle 3
The right to development must be fulfilled so as to equitably meet developmental and environmental needs of present and future generations.

Principle 4
In order to achieve sustainable development, environmental protection shall constitute an integral part of the development process and cannot be considered in isolation from it.

Principle 5
All States and all people shall cooperate in the essential task of eradicating poverty as an indispensable requirement for sustainable development, in order to decrease the disparities in standards of living and better meet the needs of the majority of the people of the world.

Principle 6
The special situation and needs of developing countries, particularly the least developed and those most environmentally vulnerable, shall be given special priority. International actions in the field of environment and development should also address the interests and needs of all countries.

Principle 7
States shall cooperate in a spirit of global partnership to conserve, protect and restore the health and integrity of the Earth's ecosys-

tem. In view of the different contributions to global environmental degradation, States have common but differentiated responsibilities. The developed countries acknowledge the responsibility that they bear in the international pursuit of sustainable development in view of the pressures their societies place on the global environment and of the technologies and financial resources they command.

Principle 8
To achieve sustainable development and a higher quality of life for all people, States should reduce and eliminate unsustainable patterns of production and consumption and promote appropriate demographic policies.

Principle 9
States should cooperate to strengthen endogenous capacity-building for sustainable development by improving scientific understanding through exchanges of scientific and technological knowledge, and by enhancing the development, adaptation, diffusion and transfer of technologies, including new and innovative technologies.

Principle 10
Environmental issues are best handled with the participation of all concerned citizens, at the relevant level. At the national level, each individual shall have appropriate access to information concerning the environment that is held by public authorities, including information on hazardous materials and activities in their communities, and the opportunity to participate in decision-making processes. States shall facilitate and encourage public awareness and participation by making information widely available. Effective access to judicial and administrative proceedings, including redress and remedy, shall be provided.

Principle 11
States shall enact effective environmental legislation. Environmental standards, management objectives and priorities should reflect the environmental and developmental context to which they apply. Standards applied by some countries may be inappropriate and of unwarranted economic and social cost to other countries, in particular developing countries.

Principle 12

States should cooperate to promote a supportive and open international economic system that would lead to economic growth and sustainable development in all countries, to better address the problems of environmental degradation. Trade policy measures for environmental purposes should not constitute a means of arbitrary or unjustifiable discrimination or a disguised restriction on international trade. Unilateral actions to deal with environmental challenges outside the jurisdiction of the importing country should be avoided. Environmental measures addressing transboundary or global environmental problems should, as far as possible, be based on an international consensus.

Principle 13

States shall develop national law regarding liability and compensation for the victims of pollution and other environmental damage. States shall also cooperate in an expeditious and more determined manner to develop further international law regarding liability and compensation for adverse effects of environmental damage caused by activities within their jurisdiction or control to areas beyond their jurisdiction.

Principle 14

States should effectively cooperate to discourage or prevent the relocation and transfer to other States of any activities and substances that cause severe environmental degradation or are found to be harmful to human health.

Principle 15

In order to protect the environment, the precautionary approach shall be widely applied by States according to their capabilities. Where there are threats of serious or irreversible damage, lack of full scientific certainty shall not be used as a reason for postponing cost-effective measures to prevent environmental degradation.

Principle 16

National authorities should endeavour to promote the internalization of environmental costs and the use of economic instruments, taking into account the approach that the polluter should, in principle, bear the cost of pollution, with due regard to the public interest and without distorting international trade and investment.

Principle 17
Environmental impact assessment, as a national instrument, shall be undertaken for proposed activities that are likely to have a significant adverse impact on the environment and are subject to a decision of a competent national authority.

Principle 18
States shall immediately notify other States of any natural disasters or other emergencies that are likely to produce sudden harmful effects on the environment of those States. Every effort shall be made by the international community to help States so afflicted.

Principle 19
States shall provide prior and timely notification and relevant information to potentially affected States on activities that may have a significant adverse transboundary environmental effect and shall consult with those States at an early stage and in good faith.

***

Principle 22
Indigenous people and their communities, and other local communities, have a vital role in environmental management and development because of their knowledge and traditional practices. States should recognize and duly support their identity, culture and interests and enable their effective participation in the achievement of sustainable development.

Principle 23
The environment and natural resources of people under oppression, domination and occupation shall be protected.

Principle 24
Warfare is inherently destructive of sustainable development. States shall therefore respect international law providing protection for the environment in times of armed conflict and cooperate in its further development, as necessary.

Principle 25
Peace, development and environmental protection are interdependent and indivisible.

Source: *Rio Declaration on Environment and Development,* June 14, 1992, U.N. Doc. A/CONF.151/26 (vol. I), 31 I.L.M. 874 (1992).

Stockholm Declaration

The 1972 Stockholm Conference on the Human Environment was a revolutionary event. Just as the Trail Smelter case established, even though it did not originate, the idea of state responsibility for transboundary environmental harm as a norm of international law, the Stockholm Conference established (while again not originating) the idea of international environmental law as a distinct subcategory within international law. Just as in the post-Nuremberg world it had become accepted that human rights was a legitimate concern of international law, in the post-Stockholm world it became accepted that environmental protection was also such a legitimate concern.

The entire Stockholm Declaration is widely available; simply typing the words "Stockholm Declaration" into any search engine will turn up numerous copies. One of the results of the Stockholm Conference was the creation of the United Nations Environment Programme (UNEP), and the full text of the Stockholm Declaration can be found on the UNEP website.

Principles 2, 11, and 21, reproduced here, set forth concepts much discussed in international environmental law. Principle 2 sets forth the idea of intergenerational equity, while Principle 11 expresses the tension between environmental and developmental interests, leading to the difficulty of achieving sustainable development. Neither of these two principles can yet be considered as stating rule of customary international law, however. While there are some theorists and judges (including, for example, Judge Weeramantry in his separate opinion in the Gabčíkovo-Nagymaros case) who are willing to argue that sustainable development, at least, has attained such normative status, this argument does not appear to be supported by the practice of states undertaken out of a sense of legal obligation.

Principle 21, on the other hand, with its statements on state responsibility and sovereignty in exploiting natural resources, is universally recognized as stating a rule of customary international law.

Principle 2

The natural resources of the earth, including the air, water, land, flora and fauna and especially representative samples of natural ecosystems, must be safeguarded for the benefit of present and future generations through careful planning or management, as appropriate.

***

Principle 11

The environmental policies of all States should enhance and not adversely affect the present or future development potential of developing countries, nor should they hamper the attainment of better living conditions for all, and appropriate steps should be taken by States and international organizations with a view to reaching agreement on meeting the possible national and international economic consequences resulting from the application of environmental measures.

***

Principle 21

States have, in accordance with the Charter of the United Nations and the principles of international law, the sovereign right to exploit their own resources pursuant to their own environmental policies, and the responsibility to ensure that activities within their jurisdiction or control do not cause damage to the environment of other States or of areas beyond the limits of national jurisdiction.

Source: *Report of the United Nations Stockholm Conference on the Human Environment*, U.N. Doc. A/CONF.48/14/Rev. 1 (1973), 11 I.L.M. 1416 (1972).

U.S. Cases

Beanal v. Freeport-McMoRan

The long-running litigation between Tom Beanal and the Louisiana-based mining corporation Freeport-McMoRan represents an attempt to bring the environmentally destructive actions of U.S. parties committed outside the United States under the control of the U.S. legal system. The approach in Beanal differs from that in Massey, an excerpt from which follows the Beanal excerpt below. In Massey, the plaintiffs sought the application of a federal statute, the National Environmental Policy Act, to the actions of a U.S. party outside the United States. In Beanal the plaintiff seeks the application of international environmental law to such actions. The mechanism by which the plaintiff sought to do this is an ancient law, over two centuries old, called the Alien Tort Claims Act. The Act grants U.S. federal courts jurisdiction over "any civil action by an alien for a tort only, committed in violation of the law of nations or a treaty of the United States." In other words, in order for the federal dis-

trict court in Louisiana to have jurisdiction over Beanal's claim, Beanal would have to establish that the actions of which he was accusing Freeport-McMoRan constituted a violation of international law. He would not have to prove, at the jurisdictional stage, that Freeport-McMoRan had in fact done those actions; that would be done later, at a trial on the merits of the case. But, if in fact the actions complained of did not violate the law of nations or any treaty of the United States, there would be no need to proceed to trial.

In the excerpt below, the U.S. District Court for the Eastern District of Louisiana considers whether customary international environmental law—the law of nations on this topic—does in fact impose obligations that were violated by the actions allegedly committed by Freeport-McMoRan.

Note: In this case and the one that follows it, many case cites—references to other cases discussed by the court or upon which the court relies—have been omitted to enhance readability.

DUVAL, District Judge

Before the court is a motion to dismiss Plaintiff Tom Beanal's ("Beanal") claims against Freeport-McMoRan, Inc. and Freeport-McMoRan Copper & Gold, Inc. (collectively "Freeport"), which motion was heard with oral argument on October 23, 1996. Having reviewed the pleadings, the memoranda, and the applicable law, the court finds . . . . Plaintiff has failed to state a claim for an environmental tort in violation of the law of nations.

The Parties

Plaintiff Tom Beanal ("Beanal") is a resident of Tamika, Irian Jaya within the Republic of Indonesia. He is a leader of the Amungme Tribal Counsel of Lambaga Adat Suku Amungme (LEMASA). He filed suit against Freeport on April 29, 1996, individually and on behalf of all other similarly situated. Plaintiff filed his first amended complaint on May 16, 1996. Since no class has been certified, Beanal is the lone plaintiff at this stage.

Defendants Freeport-McMoRan, Inc. and Freeport-McMoRan Copper and Gold, Inc. are Delaware corporations headquartered in New Orleans, Louisiana. Freeport owns an Indonesia-based subsidiary named P. T. Freeport Indonesia ("PT-FI").

Freeport operates the "Grasberg Mine," an open pit copper, gold and silver mine situated in the Jayawijaya Mountain in Irian Jaya, Indonesia. The mine allegedly encompasses approximately 26,400 square kilometers.

***

C. ENVIRONMENTAL CLAIMS

Freeport contends that Plaintiff's environmental claims should be dismissed on at least five different grounds; the court reaches only the first. Plaintiff has failed to state a claim for environmental violations upon which relief can be granted . . . because Freeport's alleged environmental practices do not appear to have violated the law of nations.

As set forth in the complaint, Plaintiff alleges that Freeport's mining operations and drainage practices have resulted in environmental destruction with human costs to the indigenous people. The mine itself has hollowed several mountains, re-routed rivers, stripped forests, and increased toxic and non-toxic materials and metals in the river system. Amended Complaint ¶29. Another culprit is discharged water containing tailings from Freeport's mining operations, for it is from this discharge that a stream of environmental and human problems flow, including:

(1) Pollution, disruption and alteration of natural waterways leading to deforestation, *Id.* ¶26;

(2) Health safety hazards and starvation, *Id.* ¶27;

(3) Degradation of surface and ground water from tailings and solid hazardous waste, *Id.* ¶28.

Beanal alleges that the tailings drainage is mismanaged. Beanal further alleges that acid mine drainage is equally devastating, due to resulting sulfide oxidation and leaching, and also inadequately managed. In summary, Plaintiff complains:

Plaintiffs specifically allege that defendant corporations have failed to engage in a zero waste policy, unacceptable enclosed waste management system, have failed to maximize environmental rehabilitation, have failed to engage in an appropriate acid leachate control policy, have failed to adequately monitor the destruction of the natural resources of Irian Jaya and have disregarded and breached its international duty to protect one of the last great natural rain forests and alpine areas in the world. *Id.* ¶40.

The court next determines whether any of these allegations, if true, amount to a violation of the law of nations. Plaintiff has not alleged that Freeport violated a specific treaty provision. As discussed above, in order to state a claim for violation of the law of nations under §1350 [the Alien Tort Claims Act], plaintiff must establish the existence of a cognizable international tort. "These international torts, violations of current customary international law, are characterized by universal consensus in the international community as to their binding status and their content. That is, they are universal, definable, and obligatory international norms." To determine whether such a norm exists, the court may consider the works of jurists, general usage and practice of nations and judicial decisions recognizing and enforcing that law. In this instance, the court reviewed case law, the Restatement, and a recent treatise on international environmental law. Having done so, the court discerns no claim of action against Freeport based on an international tort.

As a preliminary matter, courts have recognized that §1350 may be applicable to international environmental torts. *See Aguinda v. Texaco, Inc.; Amlon Metals, Inc. v. FMC Corp.* Neither of these cases, however, found a cause of action for environmental torts in violation of the law of nations. *Aguinda* referenced the possible application of §1350 for environmental practices "which might violate international law." That suit was subsequently dismissed on grounds of comity, forum non-conveniens, and failure to join a necessary party. *Amlon* involved the shipment of allegedly hazardous copper residue to a purchaser in England for metallic reclamation purposes. Among its claims, the purchaser sought recovery in tort under the Alien Tort Statute. The court rejected plaintiff's reliance on the Stockholm principles to support a cause of action under the §1350 because "those Principles do not set forth any specific proscriptions, but rather refer only in a general sense to the responsibility of nations to insure that activities within their jurisdiction do not cause damage to the environment beyond their borders." This point is well taken with respect to Beanal's complaint.

Beanal has failed to articulate a violation of the international law. Plaintiff states that the allegations support a cause of action based on three international environmental law principles: (1) the Polluter Pays Principle; (2) the Precautionary Principle; and (3) the Proximity Principle. None of the three rises to the level an international tort. Principles of International Environmental Law I: Frameworks, Standards and Implementation 183–18 (Phillipe Sands ed., 1995) (hereinafter "Sands"). Sands includes the three principles mentioned by Plaintiff in a list of general rules and principles "which have broad, if not necessarily universal, support and are frequently endorsed in practice." *Id.* at 183. Also listed are (1) the good-neighborliness and international co-operation principle and (2) the following rule, regarded the cornerstone of international environmental law: "[T]he obligation reflected in Principle 21 of the Stockholm Declaration and Principle 2 of the Rio Declaration, namely that states have sovereignty over their natural resources and the responsibility not to cause environmental damage." *Id.* Sands concludes:

> Of these general principles and rules only Principle 21/Principle 2 and the good neighborliness/international co-operation principle are sufficiently substantive at this time to be capable of establishing the basis of an international cause of action; that is to say, to give rise to an international customary legal obligation the violation of which would give rise to a legal remedy. The status and effect of the others remains inconclusive, although they may bind as treaty obligations or, in limited circumstances, as customary obligations. *Id.*

The three principles relied on by Plaintiff, standing alone, do not constitute international torts for which there is universal consensus in the international community as to their binding status and their content. More to the point, those principles apply to "members of the international community" rather than non-state corporations. Plaintiff alleges that Freeport's environmental practices reflect corporate decisions, rather than state practices. A non-state corporation could be bound to such principles by treaty, but not as a matter of international customary law. Consistent with this conclusion, the Restatement mentions only state obligations and liability in the area of environmental law.

In sum, Beanal has failed to allege an international environmental tort. The court dismisses Beanal's environmental claims for failure to state a cause of action for violation of international environmental law. Beanal has failed to articulate a substantive claim. In addition, Beanal alleged no facts that would establish, if proven, that Freeport's environmental practices constitute state action. Even assuming for the purposes of this motion that Beanal's allegations are true, Freeport's alleged policies are corporate policies only and, however destructive, do not constitute torts in violation of the law of nations. Having so concluded, the court finds it unnecessary to rule on Freeport's remaining defenses to the environmental allegations lodged against it.

Source: *Beanal v. Freeport-McMoRan, Inc.,* 969 F. Supp. 362 (E.D. La. 1997)

Environmental Defense Fund, Inc. v. Massey

A comprehensive body of environmental law, for the most part uniformly and fairly enforced, has given a high level of protection to the natural environment within the United States. But Americans also engage in environmentally harmful activities outside the United States; for the most part, these activities are beyond the reach of U.S. environmental laws.

Extraterritorial application of domestic environmental laws—that is, applications of those laws to foreign activities of a country's citizens —has long been a goal of developed-world environmental activists. Developed countries have been accused, by their own activists as well as by developing countries, of exporting their environmental problems to developing countries whose environmental laws are less strict or, as is often the case, are less strictly enforced. Extraterritorial application would help to address this problem; corporations causing environmental harm in India or Malawi could be fined at their headquarters in London or New York.

The plaintiffs in Massey sought the application of a federal statute, the National Environmental Policy Act (NEPA) to activities carried out by the National Science Foundation (NSF), an agency of the U.S. government. NEPA requires U.S. government agencies to prepare an environmental impact statement (EIS) for potentially environmentally significant actions within the United States. An Executive Order (a rule created by the President rather than by Congress) imposes a less-stringent environmental analysis requirement for actions affecting the global commons (areas outside the jurisdiction of any state) and, in some circumstances, for actions affecting the territory of foreign states.

Green activists hoped that a decision applying NEPA to the NSF's activities would open the door to general extraterritorial application of NEPA. They were to be disappointed. A federal trial court held that NEPA was not even applicable to NSF's activities in Antarctica. The federal appellate court for the District of Columbia reversed, but its opinion focused on the special nature of Antarctica, apparently foreclosing the possibility that NEPA might also be applied to actions in foreign countries. The appellate court's reasoning in applying NEPA to the NSF's action in Antarctica, as can be seen in the excerpt below, rested on Antarctica's unique legal status as a part of the global commons over which the United States, as a practical matter, exercises a great deal of control. The same reasoning might be applied to some U.S. actions in outer space or, less easily, to some activities on the high seas, but not to actions within the territories of foreign sovereigns.

Note that the court rejected the NSF's argument that the Antarctic Environment Protocol "would, if adopted by all the proposed signatories, conflict with the procedural requirements adopted by Congress for the decision-making of federal agencies under NEPA." The court gave two reasons for finding the NSF's argument unpersuasive; one was that the Antarctic Environment Protocol had not entered into force. The protocol is now in force, however; it entered into force in 1998. If the NSF is still applicable to activities in Antarctica, it must be because the court's second reason for rejecting the NSF's argument is in itself sufficient to support that application.

MIKVA, Chief Judge:

The Environmental Defense Fund ("EDF") appeals the district court's order dismissing its action seeking declaratory and injunctive relief under the National Environmental Policy Act ("NEPA"). EDF alleges that the National Science Foundation ("NSF") violated NEPA by failing to prepare an environmental impact statement ("EIS") in accordance with Section 102(2)(C) before going forward with plans to incinerate food wastes in Antarctica . . . .

***

I. As both parties readily acknowledge, Antarctica is not only a unique continent, but somewhat of an international anomaly. Antarctica is the only continent on earth which has never been, and is not now, subject to the sovereign rule of any nation. Since entry into force of the Antarctic Treaty in 1961, the United States and 39 other nations have agreed not to assert any territorial claims to the continent or to establish rights of sovereignty there. See *The Antarctica Treaty*, 12 U.S.T. 794 (Dec. 1, 1959). Hence, Antarctica is generally considered to be a "global common" and frequently analogized to outer space.

Under the auspices of the United States Antarctica Program, NSF operates the McMurdo Station research facility in Antarctica. McMurdo Station is one of three year-round installations that the United States has established in Antarctica, and over which NSF exercises exclusive control. All of the installations serve as platforms or logistic centers for U.S. scientific research; McMurdo Station is the largest of the three, with more than 100 buildings and a summer population of approximately 1,200.

Over the years, NSF has burned food wastes at McMurdo Station in an open landfill as a means of disposal. In early 1991, NSF decided to improve its environmental practices in Antarctica by halting its practice of burning food wastes in the open by October, 1991. After discovering asbestos in the landfill, however, NSF decided to cease open burning in the landfill even earlier, and to develop quickly an alternative plan for disposal of its food waste. NSF stored the waste at McMurdo Station from February, 1991 to July, 1991, but subsequently decided to resume incineration in an "interim incinerator" until a state-of-the-art incinerator could be delivered to McMurdo Station. EDF contends that the planned incineration may produce highly toxic pollutants which could be hazardous to the environment, and that NSF failed to consider fully the consequences of its decision to resume incineration as required by the decision-making process established by NEPA.

Section 102(2)(C) of NEPA requires "all federal agencies" to prepare an EIS in connection with any proposal for a "major action significantly affecting the quality of the human environment." The EIS requirement, along with the many other provisions in the statute, is designed to "promote efforts which will prevent or eliminate damage to the environment and biosphere." Following the passage of NEPA, NSF promulgated regulations applying the EIS requirement to its decisions regarding proposed actions in Antarctica. Since the issuance of Executive Order 12114, however, NSF has contended that proposed action affecting the environment in Antarctica is governed by the Executive Order, not NEPA. Executive Order 12114 declares that federal agencies are required to prepare environmental analyses for "major Federal actions significantly affecting the environment of the global commons outside the jurisdiction of any nation (e.g., the oceans or Antarctica)." E.O. 12114 § 2–3(a). According to the Executive Order, major federal actions significantly affecting the environment of foreign countries may also require environmental analyses under certain circumstances. *Id.* Although the procedural requirements imposed by the Executive Order are analogous to those under NEPA, the Executive Order does not provide a cause of action to a plaintiff seeking agency compliance with the EIS requirement. The Executive Order explicitly states that the requirements contained therein are "solely for the purpose of establishing internal procedures for Federal agencies . . . and nothing in [the Order] shall be construed to create a cause of action."

E.O. 12114 § 3–1. Thus, what is at stake in this litigation is whether a federal agency may decide to take actions significantly affecting the human environment in Antarctica without complying with NEPA and without being subject to judicial review.

II. A. The Presumption Against Extraterritoriality. As the district court correctly noted, the Supreme Court recently reaffirmed the general presumption against the extraterritorial application of statutes in *Aramco*[.] Extraterritoriality is essentially, and in common sense, a jurisdictional concept concerning the authority of a nation to adjudicate the rights of particular parties and to establish the norms of conduct applicable to events or persons outside its borders. More specifically, the extraterritoriality principle provides that "[r]ules of the United States statutory law, whether prescribed by federal or state authority, apply only to conduct occurring within, or having effect within, the territory of the United States." As stated by the Supreme Court in *Aramco*, the primary purpose of this presumption against extraterritoriality is "to protect against the unintended clashes between our laws and those of other nations which could result in international discord."

***

[T]he presumption against extraterritoriality is not applicable when the conduct regulated by the government occurs within the United States. By definition, an extraterritorial application of a statute involves the regulation of conduct beyond U.S. borders. Even where the significant effects of the regulated conduct are felt outside U.S. borders, the statute itself does not present a problem of extraterritoriality, so long as the conduct which Congress seeks to regulate occurs largely within the United States . . . .

[W]e conclude that this case does not present an issue of extraterritoriality.

***

C. The Unique Status of Antarctica. Antarctica's unique status in the international arena further supports our conclusion that this case does not implicate the presumption against extraterritoriality. The Supreme Court explicitly stated in *Aramco* that when applying the presumption against extraterritoriality, courts should look to see if there is any indication that Congress intended to extend the statute's coverage "beyond places over which the United States has sovereignty or some measure of legislative control."

Thus, where the U.S. has some real measure of legislative control over the region at issue, the presumption against extraterritoriality is much weaker. And where there is no potential for conflict "between our laws and those of other nations," the purpose behind the presumption is eviscerated, and the presumption against extraterritoriality applies with significantly less force. Indeed, it was the general understanding that Antarctica "is not a foreign country," but rather a continent that is most frequently analogized to outer space, that led this Court to conclude in *Beattie v. United States*, that the presumption against extraterritoriality should not apply to cases arising in Antarctica. The *Beattie* Court noted that Antarctica is not a "country" at all, as it has no sovereign, and stated that "to the extent that there is any assertion of governmental authority in Antarctica, it appears to be predominately that of the United States."

Even aside from this Court's holding in *Beattie*, it cannot be seriously suggested that the United States lacks some real measure of legislative control over Antarctica. The United States controls all air transportation to Antarctica and conducts all search and rescue operations there. Moreover, the United States has exclusive legislative control over McMurdo Station and the other research installations established there by the United States Antarctica Program. This legislative control, taken together with the status of Antarctica as a sovereignless continent, compels the conclusion that the presumption against extraterritoriality is particularly inappropriate under the circumstances presented in this case. As stated aptly by a State Department official in congressional testimony shortly following the enactment of NEPA, application of [NEPA] to actions occurring outside the jurisdiction of any State, including the United States, would not conflict with the primary purpose underlying this venerable rule of interpretation—to avoid ill-will and conflict between nations arising out of one nation's encroachments upon another's sovereignty. . . . There are at least three general areas: the high seas, outer space, and Antarctica. See Memorandum of C. Herter, Special Assistant to the Secretary of State for Environmental Affairs . . . [hereinafter cited as State Dept. Memo].

While the State Department memo is hardly a part of appropriate legislative history, and is not entitled to any particular deference, the memo does reflect the general understanding by those intimately involved in the creation and execution of U.S. foreign policy that the global commons, including Antarctica, do not present the challenges inherent in relations between sovereign nations. Thus, in a sovereignless region like Antarctica, where the United States has exercised a great measure of legislative control, the presumption against extraterritoriality has little relevance and a dubious basis for its application.

D. Foreign Policy Considerations. Although NSF concedes that NEPA only seeks to regulate the decision-making process of federal agencies, and that this case does not present a conflict between U.S. and foreign sovereign law, NSF still contends that the presumption against extraterritoriality controls this case. In particular, NSF argues that the EIS requirement will interfere with U.S. efforts to work cooperatively with other nations toward solutions to environmental problems in Antarctica. In NSF's view, joint research and cooperative environmental assessment would be "placed at risk of NEPA injunctions, making the U.S. a doubtful partner for future international cooperation in Antarctica." Appellee's Brief at 45.

NSF also argues that the Protocol on Environmental Protection to the Antarctic Treaty, which was adopted and opened for signature on October 4, 1991, would, if adopted by all the proposed signatories, conflict with the procedural requirements adopted by Congress for the decision-making of federal agencies under NEPA. See Protocol on Environmental Protection to the Antarctic Treaty, with Annexes, XI ATSCM, reprinted in 30 Int'l Legal Materials 1461 (1991). According to NSF, since NEPA requires the preparation of an EIS for actions with potentially "significant" impacts, while the Protocol requires an environmental analysis even for actions with "minor or transitory" impacts on the Antarctic environment, the two regulatory schemes are incompatible and will result in international discord.

We find these arguments unpersuasive. First, it should be noted that the Protocol is not in effect in any form and is years away from ratification by the United States and all 26 signatories. Second, we are unable to comprehend the difficulty presented by the two standards of review. It is clear that NSF will have to perform fewer studies under NEPA than under the Protocol, and where an EIS is required under NEPA, it would not strain a researcher's intellect to indicate in a single document how the environmental impact of the proposed action is more than "minor" and also more than "significant."

More importantly, we are not convinced that NSF's ability to cooperate with other nations in Antarctica in accordance with U.S. foreign policy will be hampered by NEPA injunctions . . . .

***

E. NEPA's Plain Language and Interpretation. NSF's final argument is that even if the presumption against extraterritoriality does not apply to this case, the plain language of Section 102(2)(C) precludes its application to NSF's decision-making regarding proposed agency action in Antarctica. We read the plain language differently. Section 102(2)(C), on its face, is clearly not limited to actions of federal agencies that have significant environmental effects within U.S. borders. This Court has repeatedly taken note of the sweeping scope of NEPA and the EIS requirement. Far from employing limiting language, Section 2 states that NEPA is intended to "encourage productive and enjoyable harmony between man and his environment " as well as to "promote efforts which will prevent or eliminate damage to the environment and biosphere." Clearly, Congress painted with a far greater brush than NSF is willing to apply[;] "there appears to have been a conscious effort to avoid the use of restrictive or limiting terminology."

***

Conclusion
Applying the presumption against extraterritoriality here would result in a federal agency being allowed to undertake actions significantly affecting the human environment in Antarctica, an area over which the United States has substantial interest and authority, without ever being held accountable for its failure to comply with the decision-making procedures instituted by Congress—even though such accountability, if it was enforced, would result in no conflict with foreign law or threat to foreign policy. NSF has provided no support for its proposition that conduct occurring within the United States is rendered exempt from otherwise applicable statutes merely because the effects of its compliance would be felt in the global commons. We therefore reverse the district court's decision, and remand for a determination of whether the environmental analyses performed by NSF, prior to its decision to resume incineration, failed to comply with Section 102(2)(C) of NEPA.

We find it important to note, however, that we do not decide today how NEPA might apply to actions in a case involving an actual foreign sovereign or how other U.S. statutes might apply to Antarctica. We only hold that the alleged failure of NSF to comply with NEPA before resuming incineration in Antarctica does not implicate the presumption against extraterritoriality.

Reversed and remanded.

Source: *Environmental Defense Fund, Inc. v. Massey*, 986 F.2d 528 (D.C. Cir. 1993).

7

Directory of Organizations, Associations, and Agencies

N umerous organizations take an interest in international environmental law. These organizations can be loosely divided into two groups, intergovernmental organizations (IGOs) and nongovernmental organizations (NGOs), although some groups may not fit neatly into either category. The IUCN and its six commissions, for example, include both national governments and NGOs as members.

The list provided here is representative rather than comprehensive; it includes general-purpose IGOs, specialized IGOs such as treaty secretariats, activist NGOs such as Greenpeace, and moderate think tanks such as Resources for the Future. The organizations chosen for inclusion here have been chosen not only for their representative quality, but for the usefulness of their websites. Most researchers in international environmental law will probably do most of their research online, as many materials will otherwise be unavailable in even the largest international law collections. The websites listed here provide, either on their own pages or via links, nearly all of the primary source documents (treaties, international judicial opinions, and aspirational documents) referred to in this book, as well as a tremendous wealth of secondary source material (books, pamphlets, and press releases).

The telephone numbers listed here include area codes in parentheses (for U.S. and Canadian telephone numbers) or country codes, separated from city codes and the remainder of the telephone number with a hyphen (for overseas telephone num-

251

bers). The international dialing prefix 011 must be used when calling most overseas numbers from most U.S. telephones.

Intergovernmental Organizations

As the number of general-purpose environmental IGOs continues to increase, the difficulty of keeping track of them increases as well. Luckily, most of these IGOs are in some way connected to the United Nations, and those that are not often make an effort to communicate their activities to the U.N., which in turn has delegated the task of keeping track of the world's environmental IGOs and coordinating their activities to the United Nations Environment Programme (UNEP). The UNEP website is thus a logical starting point for research, although the IUCN and International Environment House websites also make good starting points. Eventually the ECOLEX database may become the one-stop shop for environmental IGO documents and information, but it is still in its early stages of development.

Commission on Ecosystem Management of the IUCN
Website: http://www.iucn.org/themes/cem/cem/index.htm

The Commission on Ecosystem Management (CEM) is concerned more with the scientific dimension of international environmental problems than with law or policy issues; these issues are addressed by other IUCN commissions. The CEM gathers and publishes scientific data so that policy decisions can be made on an informed basis. The CEM website offers CEM publications for download, as well as hard copies for sale.

Commission on Education and
   Communication of the IUCN
Rue Mauverney 28
Gland 1196
Switzerland
Tel: 41-22-999-0283
Fax: 41-22-999-0025
E-mail: cec@iucn.org
Website: http://www.iucn.org/themes/cec/cec/
   home_page.htm

The Commission on Education and Communication (CEC) website provides links to a number of useful documents and sites, including sites for the Convention on Biological Diversity, the Convention on International Trade in Endangered Species, the Convention on the Conservation of Migratory Species of Wild Animals, the Ramsar Convention on Wetlands, the United Nations Convention to Combat Desertification, the United Nations Framework Convention on Climate Change and its Kyoto Protocol, and other international agreements, as well as to documents from the Rio and Johannesburg summits.

Commission on Environmental,
  Economic and Social Policy of the IUCN
c/o CENESTA: Centre for Sustainable Development
5 Lakpour Lane, Suite 24
IR-16936 Tehran, Iran
Tel: 98-21-295-4217
Fax: (253) 322-8599
E-mail: ceesp@iucn.org
Website: http://www.iucn.org/themes/ceesp

The IUCN Commission on Environmental, Economic and Social Policy (CEESP) gathers information and provides advice on economic, social, and cultural factors affecting environmental policy and the environment. It is especially concerned with the problems faced by developing countries and indigenous peoples. The CEESP website provides numerous links to other intergovernmental and nongovernmental organizations.

ECOLEX
Website: http://www.ecolex.org/ecolex/en/info.php?
  language=en
E-mail: feedback@elc.iucn.org

ECOLEX is a searchable online database of international environmental law information. It can be searched for treaties, domestic legislation and cases, and scholarly works. The database is still in development; it is operated jointly by the Food and Agriculture Organization, IUCN, and the United Nations Environment Programme, and merges databases previously created and maintained independently by these three organizations. Its eventual goal is to provide a single point of access for all information on international environmental law.

Environmental Law Programme of the IUCN
IUCN-Environmental Law Centre
Godesberger Allee 108–112
53175 Bonn
Germany
Tel: 49-228-269-2235
Fax: 49-228-269-2270
Email: Anni.Lukacs@iucn.org
Website: http://www.iucn.org/themes/law/

The Environmental Law Programme (ELP) includes, but is not limited to, the IUCN Commission on Environmental Law (CEL). Like the other IUCN commissions, the CEL is composed mostly of volunteers from around the world who dedicate their efforts to gathering and publishing information (in this case on international environmental law). The ELP also includes the Environmental Law Centre (ELC), which employs over a dozen specialists in environmental law, policy, and information resources in Bonn, as well as an international network of environmental lawyers. According to the ELP's mission statement, the ELP's mission is "to lay the strongest possible legal foundation at the international, regional, and national levels for environmental conservation in the context of sustainable development." The ELP's website provides links to numerous environmental treaty secretariats and other international and intergovernmental organizations and courts. It also provides links to other research tools, including the ECOLEX database project, and a directory of nongovernmental environmental organizations, with many links. The ELP's Environmental Policy and Law Papers can also be downloaded from the site.

European Commission Environment Directorate-General
Information Centre
Office: BU-9 01/11
B-1049 Brussels
Belgium
Fax: 32-2-299-6198
Website: http://www.europa.eu.int/comm/environment/
   index_en.htm

Most Americans have little or no understanding of the European Union (EU) and its laws. One effect of EU law is immediately obvious, however. In the period since the EU instituted a body of supranational environmental law and brought about harmonization of national laws, the quality of air and water in the affected countries has improved considerably. The website of the Environment Directorate-General provides access to all of the necessary documents: EU legislation, cases, debates and proceedings, summaries of the principles of sustainable development and other principles of international environmental law, annual surveys on the implementation and enforcement of EU environmental law, and links to other European and international environmental sites.

European Environment Agency
Kongens Nytorv 6
DK-1050 Copenhagen K
Denmark
Tel: 45-33-36-7100
Fax: 45-33-36-7199
Website: http://www.eea.eu.int/

The European Environment Agency (EEA) was established by the European Union (EU) to support the development and implementation of sound environmental policies by providing accurate environmental information to policy makers and the public. The EEA is a source of detailed environmental information on the twenty-five EU Member States, as well as Bulgaria, Iceland, Liechtenstein, Norway, Romania, and Turkey; EEA materials are available in twenty-four languages, although not all are available in every language.

Intergovernmental Panel on Climate Change Secretariat
c/o World Meteorological Organization
7bis Avenue de la Paix
C.P. 2300, CH-1211 Geneva 2
Switzerland
Tel: 41-22-730-8208
Fax: 41-22-730-8025
E-mail: IPCC-Sec@wmo.int
Website: http://www.ipcc.ch/

UNEP and the World Meteorological Organization (WMO) established the Intergovernmental Panel on Climate Change (IPCC) in 1988; membership is open to all members of the United Nations and the WMO. The IPCC collects scientific, economic, demographic, and other data related to anthropogenic climate change. It is a clearinghouse; rather than doing its own research, it collects, organizes, and summarizes the published results of original research of others in the field.

International Court of Justice (ICJ)
Peace Palace
2517 KJ The Hague
The Netherlands
Tel: 31-70-302-2323
Tel: 31-70-364-9928
E-mail: information@icj-cij.org; mail@icj-cij.org
Website: http://www.icj-cij.org/

The ICJ is the judicial organ of the United Nations; it and its predecessor, the Permanent Court of International Justice (PCIJ), are collectively referred to as the World Court. The ICJ's website provides the text of all of the ICJ's decisions, although older decisions have been scanned as images rather than text. It also provides the court's Statute and Rules, the United Nations Charter, and the text of applications, written and oral pleadings, and various other documents related to recent cases. Quite recently the ICJ website has added decisions and other documents from the PCIJ (1922–1946), although as of 16 September 2004 this area of the website was experiencing some glitches.

International Environment House
9–15 Chemin des Anemones
CH-1219 Chatelaine, Geneva
Switzerland
Website: http://www.environmenthouse.ch

International Environment House (IEH) is neither an organization nor a house; rather, it is two office buildings in Geneva, Switzerland, housing a substantial portion of the world's international environmental law administrative bodies. Treaty organizations whose secretariats or other administrative offices are housed in the IEH include the Basel Convention, the Convention on International Trade in Endangered Species, the Rotterdam Convention on the Prior Informed Consent Procedure for Certain Hazardous Chemicals and Pesticides in International Trade, the

Stockholm Convention on Persistent Organic Pollutants, and the United Nations Convention to Combat Desertification. International Environment House also houses several UNEP offices, including UNEP Chemicals, the UNEP Economics and Trade Branch, the UNEP/GEF Project on Development of National Biosafety Frameworks, the UNEP Global Resource Information Database, the UNEP Information Unit for Conventions, the UNEP Post-Conflict Assessment Unit, and the UNEP Regional Office for Europe. It also houses other UN offices, including the Emergency Response Division and the European Office of the United Nations Development Programme, the Humanitarian Relief Unit and Liaison Office of United Nations Volunteers, the United Nations Geneva Office for Project Services, an office of the United Nations Human Settlements Programme, the United Nations Institute for Training and Research, an office of the United Nations Population Fund, the United Nations System-Wide Earthwatch, and an office of the United Nations World Food Programme. Many nongovernmental organizations also have offices in the IEH; the IEH website provides links to all of the organizations housed in the International Environment House, including the umbrella Geneva Environment Network, which in turn provides links to many other environmental organizations in Geneva.

IUCN-The World Conservation Union
Rue Mauverney 28
Gland 1196
Switzerland
Tel: 41-22-999-0000
Fax: 41-22-999-0002
mail@iucn.org
Website: http://www.iucn.org

The IUCN is among the world's oldest and most influential international environmental organizations; it was founded in 1948 as the International Union for the Protection of Nature (IUPN), and changed its name to the International Union for Conservation of Nature and Natural Resources (IUCN) in 1956, and again to IUCN-The World Conservation Union in 1990. Its membership is broad and includes both national governments and nongovernmental organizations (NGOs). It is explicitly dedicated to achieving sustainable development, and to this end compiles and publishes a great deal of environmental data. These publications are available for purchase through the IUCN website. IUCN also maintains libraries, including the online environmental law library ECOLEX, discussed under its own entry in this list. IUCN's six commissions (the Species Survival Commission, World Commission on Protected Areas, Commission on Environmental Law, Commission on Education and Communication, Commission on Environment, Economic and Social Policy, and Commission on Ecosystem Management) are also discussed under their own entries.

North American Commission for Environmental Cooperation (CEC)
393, Rue St-Jacques Ouest
Bureau 200
Montréal, Québec H2Y 1N9
Canada
Tel: (514) 350-4300
Fax: (514) 350-4314
E-mail: info@ccemtl.org
Website: http://www.cec.org/home/index.cfm?varlan=english

The CEC is a regional intergovernmental organization created by a treaty between Canada, Mexico, and the United States; it addresses environmental concerns affecting North America and in particular problems involving environmental and trade issues between the three parties to the North American Free Trade Agreement (NAFTA). The CEC website includes the text of the North American Agreement on Environmental Cooperation and NAFTA, along with links to the NAFTA Secretariat and other NAFTA sites, government environmental agency sites in all three NAFTA parties, and other international organizations' sites including the Organization of American States, the Organization for Economic Cooperation and Development, and the United Nations Environment Programme. It also provides CEC publications for download.

Organization for Economic Cooperation and Development (OECD) Environment Directorate
2, Rue André Pascal
F-75775 Paris Cedex 16

France
Tel: 33-1-45-2482
Website: http://www.oecd.org/department/0,2688,en_2649_
33713_1_1_1_1_1,00.html

The OECD, sometimes called the "rich countries' club," is an intergovernmental organization whose thirty members are mostly developed countries, along with a couple of middle-income developing countries. In 1971, shortly before the Stockholm Conference, the OECD formed an Environment Policy Committee (EPOC). Gradually other environmental bodies were added, and today EPOC supervises numerous subsidiary bodies, committees, and working groups, and participates in the OECD's Working Party on Trade and the Environment, Working Party on Agriculture and Environment, and Joint Meeting of Experts on Taxation and Environment. In addition to EPOC and its related bodies, the Environment Directorate also includes the OECD's separately funded Environment, Health and Safety Programme, the OECD Chemicals Committee, the Task Force for the Implementation of the Environmental Action Programme in Central and Eastern Europe, and the Annex I Experts' Group on Climate Change. Documents relating to and produced by all of these committees, as well as OECD legal documents, are available from the OECD website.

Organization of American States (OAS), Unit for Sustainable Development and Environment
1889 F St., N.W., 6th Floor
Washington, D.C. 20006, U.S.A.
Tel: (202) 458-3000
Website: http://www.oas.org/usde/

The thirty-five members of the OAS include all of the sovereign states in North America, South America, and the Caribbean, although Cuba has been excluded from actual participation since 1962. For the past four decades the Unit for Sustainable Development and Environment (USDE) has been charged with balancing environmental and developmental concerns in the Americas. Among the first areas dealt with by the USDE, and still an area of specialization, was the design, funding, and implementation of transboundary freshwater resource projects; more recently it has also addressed a wide variety of concerns including biodiversity, alternative energy, and the role of indigenous peoples in sustainable development and environmental protection. The USDE website provides information on employment opportunities and internships with USDE, as well as a wide variety of documents available for download from the USDE's FIDA (Foro Interamericano de Derecho Ambiental) online library, including both general and Americas-specific documents.

Species Survival Commission of IUCN
Rue Mauverney 28
Gland 1196
Switzerland
Tel: 41-22-999-0000
Fax: 41-22-999-0015
Email: ssc@iucn.org
Website: http://www.iucn.org/themes/ssc/

The Species Survival Commission (SSC) is, as its name implies, concerned with extinction. The SSC, the largest of the IUCN's six commissions, gathers and collates information on endangered species and provides this information to the IUCN, as well as to governments, intergovernmental and nongovernmental organizations, and the public; the SSC's Red List of threatened species, along with its action plans and policy guidelines, are available from the SSC website. The SSC also carries out activist work and conservation projects of its own.

**United Nations Department of Economic and Social Affairs
    Division for Sustainable Development**
Two United Nations Plaza, Room DC2–2220
New York, NY 10017, USA
Tel: (212) 963-2803
Fax: (212) 963-4260
E-mail: dsd@un.org
Website: http://www.un.org/esa/sustdev/

The Division for Sustainable Development acts as a secretariat for the United Nations Commission on Sustainable Development; it seeks to implement the sustainable development principles set out in the Rio Summit's Agenda 21, the Johannesburg Plan of Implementation, and other aspirational documents. These documents may be found on the division's website.

United Nations Environment Programme (UNEP)
Nairobi, Kenya
Website [home]: http://www.unep.org
[Environmental Law Programme]:
http://www.unep.org/DPDL/Law/
[UNEP Geneva]: http://www.unep.ch

UNEP should be the starting point for any online research project in international environmental law. UNEP, created as a result of the 1972 Stockholm Conference, oversees and coordinates the environmental activities of the various agencies of the United Nations. Its website provides access to an enormous amount of information in UNEP reports; recent UNEP reports and publications have covered topics such as an assessment of the environmental effects of the war in Iraq, worldwide mercury levels, ecotourism, and atlases of freshwater resource treaties and coral reefs. The UNEP Environmental Law Programme site also provides links to treaties, aspirational documents, and, via ECOLEX, to national environmental laws and court decisions. UNEP's website also includes sites and documents relating to special projects such as the Partnership for Development of Environmental Law and Institutions in Africa (PADELIA), an attempt to improve environmental legislation, regulation, and compliance in several African countries. The UNEP Geneva site, like the International Environment House site described above, provides links to all UNEP programs and treaty secretariats located in Geneva.

World Commission on Protected Areas of the IUCN
Rue Mauverney 28
Gland 1196
Switzerland
Tel: 41-22-999-0160
Fax: 41-22-999-0015
E-mail: wcpa@iucn.org
Website: http://www.iucn.org/themes/wcpa/

The mission statement of the World Commission on Protected Areas (WCPA) declares that the "WCPA's international mission is to promote the establishment and effective management of a world-wide representative network of terrestrial and marine protected areas, as an integral contribution to the IUCN mission." The WCPA seeks to coordinate information about protected areas, such as national parks and wildlife reserves, and to coordinate policy in order to enhance the protection afforded to these areas. To this end, the WCPA gathers and disseminates information; through its website it acts as a forum for those working in the field, and many WCOA documents are available for download from the website.

World Meteorological Organization (WMO)
7bis, Avenue de la Paix
Case Postale No. 2300
CH-1211 Geneva 2
Switzerland
Tel: 41-22-730-8111
Fax: 41-22-730-8181
E-mail: wmo@wmo.int
Website: http://www.wmo.int

The WMO is one of the world's oldest international organizations. It was founded in 1873 as the International Meteorological Organization; in 1950 it was merged into the United Nations framework as a specialized agency. As its name implies, the WMO is concerned with weather, climate, and related phenomena. Weather and climate are necessarily international; they both affect the human environment and are themselves affected by human activities. As a result, the WMO's activities are directly connected to environmental concerns. WMO environmental programs include the World Weather Watch Programme, World Climate Programme, Atmospheric Research and Environment Programme, and Hydrology and Water Resources Programme. The WMO provides advice and meteorological data relevant to treaty regimes such as the ozone and global warming regimes, as well as aid in planning for natural disasters. Its website includes WMO documents and publications and compilations of technical climatological, hydrological, and meteorological data.

Treaty Secretariats

Most multilateral environmental agreements assign their administrative functions to a secretariat; some delegate these functions to an already-existing IGO. Most of these secretariats and IGOs have websites, and these websites are generally the best readily available source of comprehensive information about the treaty, its status, and its implementation. For multilateral environmental agreements not listed here, the website of the secretariat or administrative authority can usually be located by typing the name of the treaty into a search engine.

Basel Convention (UNEP)
International Environment House
13–15 Chemin des Anemones
CH-1219 Chatelaine, Geneva
Switzerland
Tel: 41-22-917-8218
Fax: 41-22-797-3454
E-mail: sbc@unep.ch
Website: http://www.basel.int/

The Basel Convention Secretariat is charged by the Basel Convention with administering the convention, along with its protocols and amendments. The secretariat's website includes the text of the convention, the Ban Amendment, the strategic plan, and the Basel Protocol on Liability and Compensation, as well as the status of ratifications. It also includes decisions of the conference of the parties, press releases, and other useful materials.

Convention on International Trade in Endangered Species of Wild Fauna and Flora (CITES)
International Environment House
Chemin des Anémones
CH-1219 Châtelaine, Geneva
Switzerland
Tel: 41-22-917-8139
Fax: 41-22-797-3417
Email: cites@unep.ch
Website: http://www.cites.org/

The CITES regime is one of the world's oldest and most comprehensive environmental law regimes. The CITES administrative structure is complex; in addition to the Secretariat and Conference of the Parties, it includes a Standing Committee and three specialized committees: the Animals Committee, the Plants Committee, and the Nomenclature Committee. The secretariat coordinates the activities of the other administrative entities; collects, distributes, and acts as a repository for information; issues the CITES annual report and the annexes containing the lists of endangered and threatened species; provides technical assistance when needed; and in some circumstances undertakes research and issues recommendations of its own.

In addition to the CITES treaty itself, along with its appendices and ratification status, a large number of other documents are necessary to an understanding of the international regime protecting endangered species. The CITES website also provides documents produced by the twelve (so far) conferences of the parties, including resolutions and decisions. It also includes the papers of the four CITES committees and more than twenty years' worth of notifications to the parties; any interested person may sign up to receive these notifications by e-mail as they come out.

Convention on Biodiversity (CBD)
393 Saint Jacques Street, Suite 300
Montreal, Quebec H2Y 1N9
Canada
Tel: (514) 288-2220
Fax: (514) 288-6588
E-mail: secretariat@biodiv.org
Website: http://www.biodiv.org

The CBD, adopted at the 1992 Earth Summit in Rio de Janeiro, Brazil, aims to protect genetic diversity both between and among species and ecosystems. The CBD's website includes the text and ratification status of the convention itself and its Cartagena Protocol on Biosafety, as well as decisions of the Conferences of the Parties (COP) and other COP documents. Clearinghouse and documents pages provide additional links and information.

Convention to Combat Desertification
P.O. Box 260129
Haus Carstanjen
D-53153 Bonn
Germany
Tel: 49-228-815-2800
Fax: 49-228-815-2898

E-mail: secretariat@unccd.int
Website: http://www.unccd.int/main.php

Like the UNFCC Secretariat, the UNCCD Secretariat is located in Haus Carstanjen on Martin Luther King Strasse in Bonn, Germany. The Convention to Combat Desertification addresses the problem of land degradation caused by overgrazing, aquifer depletion, changes in weather patterns, and other anthropogenic and natural causes. The UNCCD Secretariat's website includes the text of the convention and its annexes, ratification status, and explanatory information about the convention, as well as technical reports on desertification from around the world and general information and publications on desertification.

Mediterranean Action Plan
48 Vassileos Konstantinou Avenue
11635 Athens
Greece
Tel: 30-210-727-3100
Fax: 30-210-725-3196
E-mail: unepmedu@unepmap.gr
Website: http://www.unepmap.gr/

The Mediterranean Action Plan (MAP) is one of many programs under the UNEP umbrella designed to protect regional seas; the MAP unites twenty-one Mediterranean countries in an attempt to prevent further environmental damage to the Mediterranean Sea and ameliorate the damage that has already occurred. The MAP website provides links to treaties, technical reports, and other documents, although download speeds are extremely slow.

Ozone Secretariat (UNEP)
United Nations Environment Programme Headquarters
Nairobi, Kenya
Fax: 254-2-062-4691
Website: http://www.unep.org/ozone/index.shtml

The Ozone Secretariat is the body established by the Vienna Convention for the Protection of the Ozone Layer and charged with administrative functions under that treaty and the Montreal Protocol on Substances that Deplete the Ozone Layer. The secretariat does not make substantive rules under the convention and its protocol; rather, it arranges the meetings and activities of the bodies that do: the conference of the parties, meetings of the parties, and various committees and working groups. The secretariat also monitors implementation of and compliance with the convention and the protocol and receives and analyzes scientific data collected by the parties.

The Ozone Secretariat website includes the full text of the convention and the protocol, and the ratification status of each, as well as the ratification status of various amendments. It also includes numerous other documents related to the functioning of the ozone regime not readily available elsewhere.

**United Nations Framework Convention
  on Climate Change (UNFCC)**
P.O. Box 260124
D-53153 Bonn
Germany
Tel: 49-228-815-1000
Fax: 49-228-815-1999
E-mail: secretariat@unfccc.int
Website: http://unfccc.int

The website of the UNFCC is an essential resource for anyone interested in global warming. It includes not only the text of the convention and its Kyoto Protocol, but also thousands of pages of documents relating to both and to the attempts to bring the Kyoto Protocol into force. For example, the "documents" section of the website includes proceedings and reports of the conferences of the parties and various other bodies; the "press" section includes news and press releases. There's also a Kyoto Thermometer showing the progress in bringing the protocol into force.

The World Bank-Environment and the World Bank
1818 H Street, N.W.
Washington, DC 20433
Tel: (202) 473-1000
Fax: (202) 477-6391
Website: http://www.worldbank.org/environment/

Its role in funding development projects in poor countries often places the World Bank at the center of conflicts, actual or perceived, between the environment and development. The World

Bank appears to have grown more environmentally conscious in recent years, and its public pronouncements reflect an ostensible commitment to environmental values, but many environmentalists continue to distrust the World Bank on the basis of its previous record. The "environment" section of the World Bank's website provides access to a large number of documents, invaluable in assessing the bank's environmental policies and performance: policy statements, project reports, announcements and press releases, and the full text of *Making Sustainable Commitments: An Environment Strategy for the World Bank.*

World Heritage Center of the United Nations Educational, Scientific and Cultural Organization (UNESCO)
1, Rue Miollis
75732 Paris Cedex 15
France
Tel: 33-1-45-68-1000
Fax: 33-1-45-67-1690
Website: http://www.unesco.org/whc/toc/toc_index.htm

UNESCO is charged with the maintenance of the list of World Heritage Sites and Biosphere Reserves and with the protection of the sites themselves. Many of these sites, from the Florida Everglades to Lake Baikal, are protected because of their environmental significance, and many are in serious danger of irreversible environmental degradation. The World Heritage Center website includes a detailed list of all of the World Heritage Sites, as well as information on which are endangered. It also includes the text of the Convention Concerning the Protection of the World Cultural and Natural Heritage, along with additional information about the convention.

Nongovernmental Organizations

Concern about the environment has the power to unite people across national and political boundaries. Of the tens of thousands of environmental NGOs in the world, there are well over a thousand that are international in scope and membership. Some work primarily through activism; others devote their energy to gathering data to be used by others. Organizations of both types are listed here, as are the Tufts Multilaterals Project (an online treaty database) and EnviroLink (a link site for environmental activist groups).

Earth Policy Institute (EPI)
1350 Connecticut Ave. NW
Washington, DC 20036
Tel: (202) 496-9290
Fax: (202) 496-9325
Email: epi@earth-policy.org

The EPI, founded by Worldwatch founder and former head Lester Brown, is dedicated to studying the relationship between ecology and economics or, as the website declares, "to providing a vision of an environmentally sustainable economy—an eco-economy— as well as a roadmap of how to get from here to there."

EnviroLink Network
P.O. Box 8102
Pittsburgh, PA 15217
Website: http://www.envirolink.org

EnviroLink is exactly what the name implies: a site providing links to other sites of environmental interest. The site organizes the links into categories, and also provides a search engine.

Fridtjof Nansen Institute
P.O. Box 326
NO-1326 Lysaker
Norway
Tel: 47-67-11-1900
Fax: 47-67-11-1910
E-mail: post@fni.no

The Fridtjof Nansen Institute is an environmental research foundation named for the Norwegian scientist, explorer, and diplomat who in 1922 received the Nobel Peace Prize for his humanitarian work in post-World War I Europe. The institute publishes a tremendous amount of valuable environmental research, but it is perhaps best known for its annual *Yearbook of International Co-operation on Environment and Development.* The yearbook is one of the best available compilations of summaries of environmental and sustainable development data and is less aggressively partisan than some similar works. The yearbook can be ordered; through a link from the institute's website or from any bookseller the table of contents is available on the site.

Friends of the Earth International (FOEI)
P.O. Box 19199
1000 GD Amsterdam
The Netherlands
Tel: 31-20-622-1369
Fax: 31-20-639-2181
Website: http://www.foei.org

FOEI is a coalition of activist environmental groups from around the world. FOEI coordinates and shares information on activist campaigns around the world; the campaigns themselves are typically, but not always, local or regional. FOEI campaigns address the full spectrum of environmental problems, from Antarctic fishing to wetlands habitat loss. The FOEI website provides information on these campaigns and publications containing policy arguments rather than legal documents; it is invaluable as a resource for information on international environmental activism.

Greenpeace International
Ottho Heldringstraat 5
1066 AZ Amsterdam
The Netherlands
Tel: 31-20-514-8150
Fax: 31-20-514-8151
Email: supporter.services@int.greenpeace.org
Website: http://www.greenpeace.org

Greenpeace, with branches in over forty countries, is one of the word's largest, best-organized, and best-known environmental activist groups. Although Greenpeace campaigns for "sustainable trade," whenever environmental and economic interests are in conflict, Greenpeace is uncompromisingly on the side of the environment. As a result of this consistency and its high profile, some developing-world governments and activists have attacked Greenpeace as a symbol of the rich-world position in the North-South conflict: wildlife before people. Greenpeace's response can be summed up in the slogan "when the last tree is cut, the last river poisoned, and the last fish dead, we will discover that we can't eat money." Greenpeace's website offers a wealth of information related to the organizations activist campaigns; Greenpeace also conducts and commissions scientific research of its own, and the results of this research are available via a link from the Greenpeace International website to the Greenpeace UK website.

International Committee of the Red Cross (ICRC)
19 Avenue de la Paix
CH 1202 Geneva
Switzerland
Tel: 41-22-734-60-01
Fax: 41-22-733-20-57
Website [general]: http://www.icrc.org
[International humanitarian law databases]: http://www.
   icrc.org/web/eng/siteeng0.nsf/iwpList2/Info_
   resources:IHL_databases?OpenDocument

The ICRC is not an environmental organization but a humanitarian organization. The two areas overlap, though, in the law of war. The ICRC is composed, by its own statute, of between fifteen and twenty-five members, all of whom must be Swiss citizens. The ICRC's International Humanitarian Law databases include not only the text and ratification status of the four Geneva conventions and their two additional protocols, but nearly 100 other treaties and international agreements as well. The complete commentaries to the Geneva conventions and protocols are included as well; these provide valuable historical background on the meaning of terms in the treaties and may not be readily available in most libraries, bookstores, or elsewhere on the web.

International Law Association (ILA)
Charles Clore House
17 Russell Square
London WC1B 5DR
United Kingdom
Tel: 44-20-7323-2978
Fax: 44-02-07-323-3580
E-mail: info@ila-hq.org
Website: http://www.ila-hq.org

The International Law Association is among the world's oldest international nongovernmental organizations (NGOs). It was founded in 1873 for the "study, elucidation and advancement of international law, public and private, the study of comparative law, the making of proposals for the solution of conflicts of law and for the unification of law, and the furthering of international understanding and goodwill." A number of its committees deal with concerns related to international environmental law, particularly the committees on international law on biotechnology, international law on sustainable development, the outer continental shelf, transnational enforcement of environmental law, and water resources law. The ILA's website includes conference reports and resolutions of these committees.

Resources for the Future (RFF)
1616 P Street, NW
Washington, DC 20036
Tel: (202) 328-5000
Fax: (202) 939-3460
Website: http://www.rff.org

RFF is a Washington, D.C., think tank founded over fifty years ago to consider problems posed by United States dependence on foreign resources. Since then its scope has expanded significantly, and it addresses a wide variety of environmental issues, although oil remains a significant concern. RFF's scholarly reports are available from the website; many may be downloaded for free, while some must be purchased.

Tufts Multilaterals Project
Edwin Ginn Library
The Fletcher School
Tufts University
Medford, MA 02155
Tel: (617) 627-3273
Fax: (617) 627-3736
E-mail: ginnref@tufts.edu

The Tufts Multilaterals Project is an online treaty database; it provides the full texts of many environmental treaties, plus general information on treaty research and links to other treaty sites.

World Resources Institute (WRI)
10 G Street, NE, Suite 800
Washington, DC 20002
Tel: (202) 729-7600

Fax: (202) 729-7610
Website: http://www.wri.org

Another Washington think tank, the WRI also conducts research on a wide variety of environmental issues. WRI's reports can be downloaded from the website, although as of September 16, 2004, a few links were broken.

Worldwatch Institute
1776 Massachusetts Ave., NW
Washington, DC 20036-1904
Tel: (202) 452-1999
Fax: (202) 296-7365
Email: worldwatch@worldwatch.org
Website: http://www.worldwatch.org/

For thirty years Worldwatch has been perhaps the world's most famous, or from some perspectives notorious, environmental research institute. Worldwatch continues to issue a steady stream of scholarly papers, as well as its well-known annual *State of the World* summary and a variety of other publications. Most older works can be downloaded for free; newer works can be purchased either as downloads or in hard copy. The free downloads cover almost every imaginable environmental topic, making the Worldwatch site a valuable research resource, even though the conclusions and methodology in some Worldwatch publications have been criticized, notably by Bjørn Lomborg in *The Skeptical Environmentalist*.

World Wildlife Fund (WWF) International
Avenue du Mont Blanc 27
1196 Gland
Switzerland
Tel: 41-22-364-9027
Fax: 41-22-364-0526
Website: http://www.panda.org

The WWF is an environmental activist and advocacy group with branches in many countries, including the United States. The WWF International office acts as a secretariat for the global network of WWF organizations. Its website offers free downloads of WWF publications.

8

Selected Print
and Nonprint Resources

A thorough grounding in international environmental law requires first an understanding of environmental science, environmental philosophy, and international law. Far more
works have been published on these three topics than can be
listed here; the books and articles listed here are intended to provide one, or perhaps a few, of many possible paths to that goal.
For example, the listed works by Edward Abbey, Rachel Carson,
and Aldo Leopold provide an introduction to three major strands
of environmental philosophy, and are essential to an understanding of the goals of the environmental movement and environmental lawmakers today. The more recent work by Bjørn Lomborg points out the disturbing truth that the most deeply held
convictions of environmentalists, however heartfelt, are not always right.

Anthony D'Amato and Kirsten Engel's *International Environmental Law Anthology* provides a comprehensive overview of the
"law" side of the problem; some of the books and nearly all of
the articles listed here deal with more specific problems of science, law, philosophy, or all three. In addition, several useful
general reference works are listed.

This bibliography, of course, is only a starting point for future research; new works on international environmental law
are constantly being published. The journals and websites
listed here, as well as the organization websites listed in Chapter 7, are likely places in which to find up-to-the-minute information.

Books (General)

Abbey, Edward. *Desert Solitaire: A Season in the Wilderness.* New York: McGraw Hill, 1968.

Desert Solitaire was Edward Abbey's first published work of nonfiction; it provided, like Rachel Carson's *Silent Spring* and Aldo Leopold's *A Sand County Almanac,* an early philosophical basis for the environmental movement. Abbey was later to become better known for novels such as *The Monkeywrench Gang,* considered by some to have inspired the ecotage and, by extension, ecoterrorist movements. In *Desert Solitaire,* an account of a summer spent as a park ranger in what is now Arches National Park, he muses on humanity's destructive encroachment on the natural environment through tourism, development, and the mining of uranium to make nuclear weapons.

Beatley, Timothy, et al. *An Introduction to Coastal Zone Management.* Washington: Island Press, 1994.

Designed for use by coastal planning professionals and students, this book addresses environmental and legal issues arising from human settlement and development in the coastal zones of the United States, with reference to the experience of other countries. Topics include the role of the government in providing preparation before and relief after hurricanes and coastal storms; background on coastal zone physical and biological systems; and ways of minimizing or ameliorating human impact on these systems.

Benvenisti, Eyal. *Sharing Transboundary Resources: International Law and Optimal Resource Use.* Cambridge, UK: Cambridge University Press, 2002.

Israeli scholar Eyal Benvenisti is an expert on international freshwater resource law. In this book, he explores the failure of states to cooperate in the use of transboundary resources, resulting in insufficient and unsustainable use. Not all states fail to cooperate, of course, and the book compares successful and unsuccessful attempts to manage transboundary resources.

Bhagwati, Jagdish. *In Defense of Globalization.* New York: Oxford University Press, 2004.

To understand the globalization debate, it is necessary to examine the arguments of both sides. Yet so many books have been published on each side of the debate that it is difficult to know where to begin. Of the numerous pro-globalization books, Professor Bhagwati's is perhaps the best; it makes its points clearly and coherently and relies on facts and reasoned argument rather than on appeal to emotional or political prejudices. Chapter 11 deals specifically with globalization and the environment. This book is best read in conjunction with Professor Stiglitz's book, described below.

Bolla, Alexander J., and Ted L. McDorman (eds.). *Comparative Asian Environmental Law Anthology.* Durham, NC: Carolina Academic Press, 1999.

Asia, particularly East Asia, is among the world's most rapidly developing regions. This rapid development has brought both environmental problems and increased willingness and resources to confront those problems. Yet the environmental laws of Asian countries are at best poorly understood in the United States. This anthology seeks to remedy that lack of understanding. It is a comparative law, rather than international law, study; that is, it looks at the domestic law of the countries involved, rather than the better-known (from a U.S. perspective) international law governing environmental relations among them.

Carson, Rachel. *Silent Spring.* Boston: Houghton Mifflin, 1962.

Rachel Carson's 1962 classic, *Silent Spring,* is widely credited with launching the modern environmental movement in the United States and in much of the rest of the world as well. Though the book was written to address a narrowly defined problem—the indiscriminate aerial spraying of pesticides, particularly DDT—it launched a nationwide and ultimately worldwide debate about environmental and other rights, conformity, gender, the role of science in society, and the relationship of the individual to the state. Rachel Carson saw only a little of the profound change her book inspired, however; she died less than two years after its publication.

D'Amato, Anthony, and Kirsten Engel (eds.). *International Environmental Law Anthology.* Cincinnati, OH: Anderson Publishing, 1996.

The editors have take excerpts from books and law review articles by seventy scholars in the field (including themselves) and arranged the excerpts in an easily accessible format. A comprehensive table of contents enables the reader to locate excerpts by topic and subtopic. The book also includes an appendix of over 200 pages of environmental agreements.

Davidson, Eric A. *You Can't Eat GNP: Economics As if Ecology Mattered.* Cambridge, MA: Perseus Publishing, 2000.

Dr. Davidson takes an alternative approach to the problem of environmental economics. His thesis is that economic measures currently in use, particularly measures such as gross domestic product and gross national product, are fundamentally flawed because they fail to take environmental values into account: They do not reflect the massive externalities involved in resource exploitation and pollution.

Folsom, Ralph H., et al. *International Trade and Investment in a Nutshell.* St. Paul, MN: West, 1996.

In order to understand the debate about trade and the environment and the effects of international trade law on the environment, it is first necessary to understand the legal regime governing international trade. *International Trade and Investment in a Nutshell,* one of West Publishing's Nutshell series of legal guides, provides that understanding. Like all of the volumes in the Nutshell series, it is designed for lawyers and law students but should be accessible to all readers. Professor Folsom and coauthors Michael Gordon and John Spanogle are also the authors of a widely used casebook on international business transactions.

Graham, Edward M. *Fighting the Wrong Enemy: Antiglobal Activists and Multinational Enterprises.* Washington, DC: Institute for International Economics, 2001.

The pro-globalization argument most often made from an environmental perspective has two steps. First, globalization reduces poverty. Second, reduction of poverty increases willingness and ability to spend money on environmental protection. Edward Graham's book makes this argument effectively.

Guruswamy, Lakshman. *International Environmental Law in a Nutshell.* St. Paul, MN: West, 1997.

International Environmental Law in a Nutshell is intended as a supplemental study aid for law students taking classes in international environmental law and for lawyers not specializing in the field who need an overview. It is accessible to nonlawyers and nonlaw students. Professor Guruswamy is also an author of a widely used textbook on international environmental law and an internationally recognized expert in the field.

Knowledge for Sustainable Development: An Insight into the Encyclopedia of Life Support Systems. 3 vols., Paris: UNESCO Publishing, 2002.

The mammoth *Encyclopedia of Life Support Systems* (EOLSS)—actually sixteen integrated online encyclopedias—gathers together knowledge from a wide variety of disciplines that may be relevant to environmental studies in general and sustainable development studies in particular. Most of the material in EOLSS is available only online, by subscription, but the three-volume set *Knowledge for Sustainable Development* contains print versions of the theme-level (that is, lengthier and more general) entries.

Leopold, Aldo. *A Sand County Almanac.* New York: Oxford University Press, 1949.

On its surface, *A Sand County Almanac* is a series of essays about nature, many of them (including all of those in Part I) concerning observations made on or around Aldo Leopold's farm in Sand County, Wisconsin. On a deeper level the essays are about the failure of the modern world, and the United States in particular, to understand the natural environment. In the book Leopold sets forth his "land ethic." Nothing that disturbs the balance of nature is right. Humanity's role in nature should be to preserve as much wild land as possible to ensure the future of all species. *A Sand County Almanac* started a quieter and slower revolution than Rachel Carson's *Silent Spring;* it has inspired a more radical form of environmentalism, one that does not put humanity at the center of environmental concerns—although the author is careful to point out that "wild things . . . had little human value until mechanization assured us of a good breakfast."

Lomborg, Bjørn. *The Skeptical Environmentalist: Measuring the Real State of the World.* Cambridge, UK: Cambridge University Press, 2001.

Danish statistician Bjørn Lomborg's exploration of environmental facts and trends has become the most controversial environmental work published in recent years. Often characterized as an anti-Green tract, Lomborg's book may also be read as an argument for a wiser allocation of environmental protection resources. An unfortunate characteristic not only of environmental debate, but of modern political debate generally is an unwillingness to do the hard work of actually looking at facts; many people seem to feel that they know what the facts are, and attempt to persuade their arguments through sheer force of argument or emotion, or with threats and bribes. Lomborg concludes that many environmental problems, including some that have a very high media profile, may not be as bad as has been feared and may in fact not be problems at all. He may well be incorrect; but attempts to prove him so will only be effective if they can use verifiable facts, rather than impassioned argument, to challenge his conclusions.

McCaffrey, Stephen C. *The Law of International Watercourses: Non-navigational Uses.* Oxford: Oxford University Press, 2001.

Professor McCaffrey is the former special rapporteur for the International Law Commission in its work on the law of non-navigational uses of transboundary watercourses and one of the world's leading experts on international water law. This monograph, written for law students, scholars, and practitioners, covers nearly all of the major multilateral agreements and international cases in the field.

McNeill, J.R. *Something New Under the Sun: An Environmental History of the Twentieth-Century World.* New York: W.W. Norton, 2000.

An understanding of the development of modern international environmental law is impossible without an understanding of the dramatic environmental changes—particularly the change in humanity's environmental role—over the course of the twentieth century. In *Something New Under the Sun,* Professor McNeill provides a detailed, accessible look at some of the preceding century's major environmental changes.

O'Riordan, Timothy, and Heather Voisey (eds.). *The Transition to Sustainability: The Politics of Agenda 21 in Europe.* London: Earthscan, 1998.

The countries of Europe have achieved a level of international legal and political integration not found elsewhere in the world. In this book, a thematically organized collection of essays, several of them by the editors, explores the incorporation of sustainable development concepts into the domestic law of several European countries, and the supranational law of the European Union.

Stiglitz, Joseph E. *Globalization and Its Discontents.* New York: W.W. Norton, 2002.

Of the numerous authors who have written about globalization, Professor Stiglitz is perhaps the best qualified. He is a Nobel laureate in economics, a former chief economist of the World Bank, and the former chair of President Clinton's Council of Economic Advisors. *Globalization and Its Discontents* is, on balance, more antiglobalization than otherwise, and is best read in conjunction with Jagdish Bhagwati's *In Defense of Globalization.* (Both authors, incidentally, are currently professors at Columbia University.) Although no portion of the book is dedicated specifically to environmental issues, these issues are discussed in Chapter 9.

Books (Reference and Directories)

Art, Henry W., ed. *The Dictionary of Ecology and Environmental Science.* New York: Henry Holt & Co., 1993.

The student of international environmental law may often encounter the use of unfamiliar technical terms. Just as a law dictionary is essential to understand legal terminology, an environmental dictionary is necessary to understand environmental and scientific terminology. If you're not completely sure you know the difference between absorption and adsorption, this dictionary can help.

Buergenthal, Thomas, and Sean D. Murphy. *Public International Law in a Nutshell, 3rd ed.* St. Paul, MN: West, 2002.

This entry in West Publishing's Nutshell series provides an overview of international law for law students and lawyers. Thomas Buergenthal is one of the world's most celebrated international law scholars and a judge on the International Court of Justice. Professor Murphy teaches at George Washington University and has represented the United States before several international tribunals.

Danner, Richard A., and Marie-Louise H. Bernal, eds. *Introduction to Foreign Legal Systems.* London: Oceana, 1994.

Doing research in a foreign legal system is a daunting task, even for those who know the language. The entries in *Introduction to Foreign Legal Systems* provide a guide to legal research in several of the world's major systems.

Fox, James R. *Dictionary of International and Comparative Law: See* Clive Parry et al., *Encyclopaedic Dictionary of International Law,* below.

Garner, Bryan A. ed. *Black's Law Dictionary, 7th ed.* St. Paul, MN: West, 1999.

Black's Law Dictionary is a standard legal reference. Just as an environmental or scientific dictionary is necessary for the understanding of technical scientific terminology, *Black's Law Dictionary* (or a similar work) is necessary for an understanding of legal terminology. Words like "delict" and "refoulement" are not likely to be recognized by most laypersons but may be encountered in international law.

Guruswamy, Lakshman D., et al. *Supplement of Basic Documents to International Environmental Law and World Order, 2nd ed.* St. Paul, MN: West, 1999.

Finding the treaties, cases, international organization documents, and other documents upon which international law is based can be difficult and time-consuming. Most are available online; a few are available in Chapter 6 of this book. Some law libraries stock encyclopedic sets of international law documents. But what most serious students of international environmental law need is a single volume that includes the text of the major documents. The *Supplement of Basic Documents to International Environmental Law and World Order* fills that need. It contains over 1,500 pages of documents and information on their status. The *Supplement* is intended as a companion to an international law textbook by Professors Gurswamy et al., but can also be purchased separately.

Parry, Clive, et al. *Encyclopaedic Dictionary of International Law.* London: Oceana, 1986.

Fox, James R. *Dictionary of International and Comparative Law, 3rd ed.* Dobbs Ferry, NY: Oceana Publications, 2003.

For more detail on, and in some cases more coverage of, terms specific to international law than is provided by *Black's Law Dictionary*, an international law dictionary is necessary. Parry's dictionary provides brief entries, typically one paragraph but sometimes longer, with citations to authorities. Fox's dictionary provides somewhat less detail, but covers a greater number of topics and is more recently updated.

Turner, Barry, ed. *The Statesman's Yearbook: The Politics, Cultures and Economies of the World, 139th ed.* Houndmills, Basingstoke, UK: Palgrave, 2001.

The Statesman's Yearbook provides information on each of the world's countries similar to, but more detailed than, the information in the country entries in the *World Almanac*, which will also suffice as a desk reference if the *Statesman's Yearbook* is unavailable.

Journals

A great many journals on international law and environmental law are published at universities across the United States, along with a handful of journals dealing specifically with international environmental law. Most of these journals are available in law libraries. Most are also available online through Westlaw (www.westlaw.com), Lexis (www.lexis.com), and HeinOnline (www.heinonline.org), although all of these services charge a fee for access. Many may also be available from the journals' own websites.

As in Chapter 7, the telephone numbers listed here include area codes in parentheses (for United States and Canadian telephone numbers) or country codes, separated from city codes and the remainder of the telephone number with a hyphen (for overseas telephone numbers). The international dialing prefix 011 must be used when calling most overseas numbers from most U.S. telephones.

Across Borders International Law Journal
Gonzaga University School of Law
P.O. Box 3528
Spokane, WA 99220
(509) 323-5792; Fax: (509) 323-5733

Albany Law Environmental Outlook
Albany Law School
80 New Scotland Ave.
Albany, NY 12208
(518) 472-5863
Email: outlook@mail.als.edu

American Journal of Comparative Law
University of California at Berkeley
Boalt Hall School of Law
Berkeley, CA 94720
(510) 643-6115

American Journal of International Law
Vanderbilt University Law School
131 21st Ave. S., Ste. 207
Nashville, TN 37203-1181
Fax: (615) 343-7979
Email: ajil@law.vanderbilt.edu

American University International Law Review
Washington College of Law
4801 Massachusetts Ave. NW, Ste. 610

Washington, DC 20016
(202) 274-4460; Fax: (202) 274-4465
Email: auilr@wcl.american.edu

Annual Survey of International and Comparative Law
Golden Gate University School of Law
536 Mission St.
San Francisco, CA 94015
(415) 442-6653; Fax: (415) 495-6756

Arizona Journal of International and Comparative Law
University of Arizona
James E. Rogers College of Law
1201 E. Speedway, P.O. Box 210176
Tucson, AZ 85721-1076
(520) 621-5593; Fax: (520) 621-9140

Berkeley Journal of International Law
University of California at Berkeley
Boalt Hall School of Law
Berkeley, CA 94720-7200
(510) 642-9759; Fax: (510) 642-9759

Boston College Environmental Affairs Law Review
Boston College Law School
885 Centre St.
Newton, MA 02459
(617) 552-8557; Fax: (617) 552-4098
Email: ealr@bc.edu

Boston College International and Comparative Law Review
Boston College Law School
885 Centre St.
Newton, MA 02459
(617) 552-8557; Fax: (617) 552-4098
Email: iclr@bc.edu

Boston University International Law Journal
Boston University School of Law
765 Commonwealth Ave.
Boston, MA 02215
(617) 353-3157; Fax: (617) 353-3077
Email: builj@bu.edu

Brooklyn Journal of International Law
Brooklyn Law School
One Boerum Pl., 1st Fl.
250 Joralemon St.
Brooklyn, NY 11201
(718) 780-7971; Fax: (718) 780-0353
Email: intl-lj@brooklaw.edu

Buffalo Environmental Law Journal
SUNY-Buffalo Law School
403 John Lord O'Brian Hall
Buffalo, NY 14260
(716) 645-7342; Fax: (716) 645-2064
Email: belj1993@hotmail.com

California Western International Law Journal
California Western School of Law
225 Cedar St.
San Diego, CA 92101
(619) 525-1475; Fax: (619) 696-9999
Email: lawreview@cwsl.edu

Cardozo Journal of International and Comparative Law
Yeshiva University
Benjamin N. Cardozo School of Law
55 Fifth Ave.
New York, NY 10003-4391
(212) 790-0264

Case Western Reserve Journal of International Law
Case Western Reserve University School of Law
Gund Hall, 11075 E. Blvd.
Cleveland, OH 44106-7148
(216) 368-3291; Fax: (216) 368-3310
Email: casejil@case.edu.26

Chicago Journal of International Law
University of Chicago Law School
1111 E. 60 St.
Chicago, IL 60637-2786
(773) 834-4464; Fax: (773) 834-3023
Email: cjil@uchicago.edu

Colorado Journal of International Environmental Law and Policy
University of Colorado School of Law
256 Fleming Law Bldg., 401 UCB
Boulder, CO 80309-0401
(303) 492-2265; Fax: (303) 492-1200
Email: cjielp@stripe.colorado.edu

Columbia Journal of Asian Law
Columbia University Law School
435 W. 116 St., Box C-10
New York, NY 10027
(212) 854-5510; Fax: (212) 854-7946
Email: jrnasian@equinox.law.columbia.edu

Columbia Journal of Environmental Law
Columbia University Law School
435 W. 116 St., Mail Code 3513
New York, NY 10027
(212) 854-1606; Fax: (212) 854-7946
Email: jrnenv@law.columbia.edu

Columbia Journal of European Law
Columbia University Law School
Parker School of Foreign and Comparative Law
435 W. 116 St.
New York, NY 10027
(212) 854-5811; Fax: (212) 854-7956
Email: jrneur@law.columbia.edu

Columbia Journal of Transnational Law
Columbia University Law School
435 W. 116 St.
New York, NY 10027
(212) 854-1604; Fax: (212) 854-5994
Email: transnational@law.columbia.edu

Connecticut Journal of International Law
University of Connecticut School of Law
65 Elizabeth St.
Hartford, CT 06105
(860) 570-5297; Fax: (860) 570-5299
Email: cjil@law.uconn.edu

Cornell International Law Journal
Cornell Law School
Myron Taylor Hall
Ithaca, NY 14853
(607) 255-9666; Fax: (607) 255-7193

Denver Journal of International Law
  and Policy
University of Denver College of Law
2255 E. Evans Ave., Ste. 235
Denver, CO 80208
(303) 871-6166; Fax: (303) 871-6378

Duke Environmental Law and Policy Forum
Duke University School of Law
Corner of Towerview Rd. and Science Dr.
Box 90364
Durham, NC 27708
(919) 613-7224; Fax: (919) 681-8460
Email: delpf@student.law.duke.edu

Duke Journal of Comparative
  and International Law
Duke University School of Law
Corner of Towerview Rd. and Science Dr.
Box 90364
Durham, NC 27708
(919) 613-7102
Email: djcil@student.law.duke.edu

East European Constitutional Review
New York University School of Law
161 Avenue of the Americas, 12th Floor
New York, NY 10013
Fax: (212) 995-4600

Ecology Law Quarterly
University of California at Berkeley
Boalt Hall School of Law
Berkeley, CA 94720
(510) 642-0457; Fax: (510) 643-9042
Email: ecologylq@law.berkeley.edu

Emory International Law Review
Emory University School of Law
Gambrell Hall, 1301 Clifton Road
Atlanta, GA 30522
(404) 727-6816; Fax: (404) 727-2203

Energy and Mineral Law Foundation
University of Kentucky Mineral Law Center
Room 21, Law Bldg.
Lexington, KY 40506
(859) 257-1293; Fax: (859) 323-1061
Email: jnrel@yahoo.com

Energy Law Journal
University of Tulsa College of Law
John Rogers Hall
3120 E. 4th Pl.
Tulsa, OK 74104-2499
(918) 631-2044

Environmental Claims Journal
Environmental Strategies Corporation
11911 Freedom Dr.
Reston, VA 22090
(703) 709-6500

The Environmental Forum
Environmental Law Institute
1616 P St. N.W., Ste. 200
Washington, DC 20036
(202) 939-3800; Fax: (202) 939-3868
Email: law@eli.org

Environmental Law
Lewis and Clark Law School
10015 Southwest Terwilliger Blvd.
Portland, OR 97219
(503) 768-6700; Fax: (503) 768-6671

Environmental Law Reporter
Environmental Law Institute
1616 P St. N.W.
Washington, DC 20036
(202) 939-3800; Fax: (202) 939-3868

Environs: Environmental Law and
  Policy Journal
University of California at Davis School of Law
400 Mrak Hall Dr.
Davis, CA 95616
(530) 754-7903; Fax: (530) 758-7804
Email: environs@ucdavis.edu

European Environmental Law Review
Extenza-Turpin, Stratton Business Park
Pegasus Dr., Biggleswade
Bedfordshire SG18 8QB, UK
Phone: 44-1-767-60-4958
Fax: 44-1-767-60-1640

European Journal of International Law
European University Institute
Villa Schifanoia, Via Boccaccio, 121
Firenze, 50133 Italy
Tel: 39-055-468-5555
Fax: 39-055-468-5517

Florida Journal of International Law
University of Florida Levin College of Law
Holland Law Ctr., 141 Bruton-Geer Hall
Gainesville, FL 32604
(352) 392-4980; Fax: (352) 392-3005
Email: fjil@law.ufl.edu

Fordham Environmental Law Journal
Fordham University School of Law
140 W. 62 St.
New York, NY 10023
(212) 636-6946; Fax: (212) 636-6963
Email: elj@fordham.edu

Fordham International Law Journal
Fordham University School of Law
140 W. 62 St.
New York, NY 10023
(212) 636-6931; Fax: (212) 636-6932
Email: ilj@fordham.edu

Georgetown International Environmental Law Review
Georgetown University Law Center
600 New Jersey Ave., N.W.
Washington, DC 20001
(202) 662-9689; Fax: (202) 662-9492
Email: gielr@law.georgetown.edu

Georgetown Journal of International Law
Georgetown University Law Center
600 New Jersey Ave., N.W.
Washington, DC 20001
(202) 662-9690; Fax: (202) 662-9491
Email: gjil@law.georgetown.edu

George Washington International Law Review
George Washington University Law School
2008 G St., N.W.
Washington, DC 20052
(202) 676-3847; Fax: (202) 676-3876
Email: gwilr@gwu.edu

Georgia Journal of International and Comparative Law
University of Georgia School of Law
Athens, GA 30602-6012
(706) 542-7289; Fax: (706) 542-5556

Harvard Environmental Law Review
Harvard Law School
Cambridge, MA 02138
(617) 495-3110; Fax: (617) 495-1110
Email: hlselr@law.harvard.edu

Harvard International Law Journal
Harvard Law School
Cambridge, MA 02138
(617) 495-3146; Fax: (617) 495-1110
Email: hlsilj@law.harvard.edu

Hastings International and Comparative Law Review
UC-Hastings College of the Law
200 McAllister St.
San Francisco, CA 94102-4978
(415) 581-8963; Fax: (415) 581-8974
Email: hiclr@uchastings.edu

Hastings West-Northwest Journal of Environmental Law and Policy
200 McAllister St.
San Francisco, CA 94102-4978
(415) 581-8966; Fax: (415) 581-8967
Email: wnw@uchastings.edu

Houston Journal of International Law
100 University of Houston Law Center
Houston, TX 77204-6060
(713) 743-2212; Fax: (713) 743-0895

ICSID Review: Foreign Investment Law Journal
International Centre for Settlement of Investment Disputes
1818 H St., N.W.
Washington, DC 20433
(202) 458-1534; Fax: (202) 522-5615

ILSA Journal of International and Comparative Law
Nova Southeastern University
Shepard Broad Law Center
3305 College Ave.
Fort Lauderdale, FL 33314
(954) 262-6026; Fax: (954) 262-3830
Email: journal@nsu.law.nova.edu

Indiana International and Comparative Law Review
Indiana University School of Law
530 W. New York St.
Indianapolis, IN 46202-3225
(317) 274-1050; Fax: (317) 274-3955
Email: iiclriul@iupui.edu

Indiana Journal of Global Legal Studies
Indiana University School of Law
211 S. Indiana Ave.
Bloomington, IN 47405-1001
(812) 855-8717; Fax: (812) 855-0555
Email: ijgls@indiana.edu

International and Comparative Law Quarterly
Oxford University Press
Great Clarendon St.
Oxford, OX2 6DP, UK

Telephone: 44-1-865-35-3907
Fax: 44-1-865-35-3485

International and Comparative Law Review
University of Miami School of Law
P. O. Box 248087
Coral Gables, FL 33124
(305) 284-1856
Email: umiclr@students.law.miami.edu

The International Lawyer
Southern Methodist University
Dedman School of Law
P.O. Box 750116
Dallas, TX 75275-0116
(214) 768-2061; Fax: (214) 768-1996

International Legal Perspectives
Lewis and Clark Law School
10015 Southwest Terwilliger Blvd.
Portland, OR 97219
(503) 768-6778; Fax: (503) 768-6671
Email: ilp@lclark.edu

Journal of Energy and Natural Resources Law
271 Regent St
London W1B 2AQ UK

Journal of Environmental Law
Oxford University Press
Great Clarendon St.
Oxford, OX2 6DP, UK
Telephone: 44-1-865-35-3907
Fax: 44-1-865-35-3485
www.jel.oupjournals.org

Journal of Environmental Law and Litigation
University of Oregon School of Law
138 Knight Law Center
1515 Agate St.
Eugene, OR 97403-1221
(541) 346-3891; Fax: (541) 346-1564
Email: jell-eic@law.uoregon.edu

Journal of International Law
Michigan State University
Detroit College of Law
209 Law College Bldg.
E. Lansing, MI 48824-1300
(517) 432-6932; Fax: (517) 432-6966
Email: dcljilp@msu.edu

Journal of International Wildlife Law and Policy
1702 Arlington Blvd.
El Cerrito, CA 94530
(650) 281-9126; Fax: (801) 838-4710

Journal of Land, Resources and Environmental Law
University of Utah, S.J. Quinney College of Law
332 S. 1400 E.
Salt Lake City, UT 84112-0730
(801) 581-3583; Fax: (801) 585-6897

Journal of Land Use and Environmental Law
Florida State University College of Law
Tallahassee, FL 32306-1601
(850) 644–4240; Fax: (850) 644-7482

Journal of Natural Resources and Environmental Law
University of Kentucky College of Law
620 S. Limestone St.
Lexington, KY 40506-0048
(859) 257-1293; Fax: (859) 323-1061
Email: jnrel@yahoo.edu

Journal of Transnational Law and Policy
Florida State University College of Law
Tallahassee, FL 32306-1601
(850) 644-0961; Fax: (850) 644-7780
Email: jtlp@law.fsu.edu

*Loyola/Los Angeles International and Comparative Law
  Review*
Loyola Law School
919 S. Albany St.
Los Angeles, CA 90015-0019

(213) 736-1405; Fax: (213) 385-6247
Email: ilr@lls.edu

Michigan Journal of International Law
University of Michigan Law School
625 S. State St.
Ann Arbor, MI 48109-1215
(734) 763-4597; Fax: (734) 764-6043

Minnesota Journal of Global Trade
University of Minnesota Law School
220 Law Center
229 19th Ave. S.
Minneapolis, MN 55455
(612) 625-6884; Fax: (612) 625-2011
Email: mnjrlglobetrade@hotmail.com

Missouri Environmental Law and Policy Review
University of Missouri School of Law
Columbia, MO 65211
Email: umclawmelpr@missouri.edu

Natural Resources Journal
University of New Mexico School of Law
MSC11 6070, 1 University of New Mexico
Albuquerque, NM 87131-0001
(505) 277-4910; Fax: (505) 277-8342
Email: nrj@law.unm.edu

New York International Law Review
New York State Bar Association
One Elk St.
Albany, NY 12207
(518) 463-3200

*New York Law School Journal of International and
    Comparative Law*
New York Law School
57 Worth St.
New York, NY 10013-2960
(212) 431-2113; Fax: (212) 966-9153
Email: ilj@nyls.edu

New York University Environmental Law Journal
New York University School of Law
110 W. Third St.
New York, NY 10012
(212) 998-6560; Fax: (212) 995-4032
Email: law.elj@nyu.edu

New York University Journal of International Law and Politics
New York University School of Law
110 W. Third St.
New York, NY 10012
(212) 998-6520; Fax: (212) 995-4032
Email: journal.international@nyu.edu

North Carolina Journal of International Law and Commercial Regulation
University of North Carolina School of Law
Van Hecke-Wettach Hall, CB 3380
Chapel Hill, NC 27599-3380
(919) 962-4402; Fax: (919) 962-4713
Email: ncilj@unc.edu

Northwestern Journal of International Law and Business
Northwestern University School of Law
357 E. Chicago Ave.
Chicago, IL 60611
(312) 503-8742; Fax: (312) 503-0132
Email: jilb@law.northwestern.edu

Ocean Development and International Law
Prof. Jon L. Jacobson, Editor-in-Chief
University of Oregon School of Law
Eugene, OR 97403-1221
(541) 346-3837; Fax: (541) 346-1564

Oil, Gas and Energy Quarterly
Matthew Bender & Company
Two Park Ave.
New York, NY 10016
(212) 448-2000

Oregon Review of International Law
University of Oregon School of Law
1515 Agate St.
Eugene, OR 97403
(541) 346-3849; Fax: (541) 346-1596
Email: oril@law.uoregon.edu

Pace Environmental Law Review
Pace University School of Law
78 North Broadway
White Plains, NY 10603
(914) 422-4116; Fax: (914) 422-4275

Pace International Law Review
Pace University School of Law
78 North Broadway
White Plains, NY 10603
(914) 422-4271; Fax: (914) 422-4275
Email: pilr@law.pace.edu

Pacific Rim Law and Policy Journal
University of Washington School of Law
William H. Gates Hall
1100 N. E. Campus Parkway
Seattle, WA 98105-6617
(206) 543-6649; Fax: (206) 685-4457
Email: pacrim@u.washington.edu

Parker School Journal of East European Law
Columbia University Law School
435 W. 116 St.
New York, NY 10027
(212) 854-2268; Fax: (212) 222-4256

Penn State Environmental Law Review
Pennsylvania State University
Dickinson School of Law
150 S. College St.
Carlisle, PA 17013-2899
(717) 240-5006; Fax: (717) 243-4443
Email: pselr@psu.edu

Penn State International Law Review
Pennsylvania State University
Dickinson School of Law
150 S. College St.
Carlisle, PA 17013-2899
(717) 240-5233; Fax: (717) 241-3511

Public Land and Resources Law Review
University of Montana School of Law
Missoula, MT 59812
(406) 243-6568; Fax: (406) 243-2576
Email: publand@selway.umt.edu

Regent Journal of International Law
Regent University School of Law
1000 Regent University Dr., 304D
Virginia Beach, VA 23464-9800
(757) 226-4101; Fax: (757) 226-4338
Email: rjil@regent.edu

San Diego International Law Journal
University of San Diego School of Law
5998 Alcala Park
San Diego, CA 92110
(619) 260-4531; Fax: (619) 260-7497
Email: sdilj@sandiego.edu

Santa Clara Journal of International Law
Santa Clara University School of Law
500 El Camino Real
Santa Clara, CA 95053-0421
(408) 554-5335
Email: scjilsubmissions@scu.edu

South Carolina Journal of International Law
and Business
University of South Carolina School of Law
Columbia, SC 29208
(803) 777-1429
Email: scjilb@law.law.sc.edu

Southeastern Environmental Law Journal
University of Carolina School of Law
Main and Greene Streets
Columbia, SC 29208
(803) 777-9329; Fax: (803) 777-5827
Email: elj@student.law.sc.edu

Stanford Environmental Law Journal
Stanford Law School
559 Nathan Abbott Way
Stanford, CA 94305-8610
(650) 725-0183; Fax: (650) 723-0202

Stanford Journal of International Law
Stanford Law School
559 Nathan Abbott Way
Stanford, CA 94305-8610
(650) 723-1375; Fax: (650) 723-0202

Suffolk Transnational Law Review
Suffolk University Law School
120 Tremont St.
Boston, MA 02108-4977
(617) 573-8180; Fax: (617) 723-5847

Syracuse Journal of International Law and Commerce
Syracuse University College of Law, Ste. 410
Syracuse, NY 13244-1030
(315) 443-2056; Fax: (315) 443-4026

Temple International and Comparative Law Journal
Temple University, Beasley School of Law
1719 North Broad St.
Philadelphia, PA 19122-1185
(215) 204-8945; Fax: (215) 204-1185
Email: ticlj@temple.edu

Texas Environmental Law Journal
University of Texas School of Law
727 E. Dean Keeton St.
Austin, TX 78705
(512) 471-0299
Email: telj@mail.law.utexas.edu

Texas International Law Journal
University of Texas School of Law
727 E. Dean Keeton St.
Austin, TX 78705
(512) 232-1277; Fax: (512) 471-6988
Email: tilj@mail.law.utexas.edu

Third World Legal Studies
International Third World Legal Studies Association
City University of New York School of Law
65-21 Main St.
Flushing, NY 11367

Toledo Journal of Great Lakes Law, Science and Policy
University of Toledo College of Law
2801 W. Bancroft St.
Toledo, OH 43606-3390
(419) 530-2876
Email: jlsp@utoledo.edu

Touro Environmental Law Journal
Touro College
Jacob D. Fuchsberg Law Center
300 Nassau Road
Huntington, NY 11743
(516) 421-2244; Fax: (516) 425-2675

Touro International Law Review
Touro College
Jacob D. Fuchsberg Law Center
300 Nassau Road
Huntington, NY 11743
(631) 421-2244; Fax: (631) 421-2675

Transnational Law and Contemporary Problems
University of Iowa College of Law
Boyd Law Bldg., Room 185
Iowa City, IA 52242-1113
(319) 335-9736; Fax: (319) 353-5817
Email: tlcp@uiowa.edu

Transnational Lawyer
University of the Pacific
McGeorge School of Law

3200 Fifth Ave.
Sacramento, CA 95817-9989
(916) 739-7133; Fax: (916) 739-7111
Email: ttl@uop.edu

Tulane Environmental Law Journal
Tulane University Law School
John Giffen Weinmann Hall
6329 Freret St.
New Orleans, LA 70118-5670
(504) 865-5309; Fax: (504) 865-6748

Tulane European and Civil Law Forum
Tulane University Law School
John Giffen Weinmann Hall
6329 Freret St.
New Orleans, LA 70118-5670
(504) 862-8839; Fax: (504) 865-6748
Email: lbecnel@law.tulane.edu

*Tulane Journal of International
and Comparative Law*
Tulane University Law School
John Giffen Weinmann Hall
6329 Freret St.
New Orleans, LA 70118-5670
(504) 862-8640; Fax: (504) 865-6748
Email: lbecnel@law.tulane.edu

*Tulsa Journal of Comparative and
International Law*
The University of Tulsa College of Law
3120 E. Fourth Pl.
Tulsa, OK 74104
(918) 631-3190; Fax: (918) 631-2194
Email: tjcil@utulsa.edu

UCLA Journal of Environmental Law and Policy
UCLA School of Law
405 Hilgard Ave., Box 951476
Los Angeles, CA 90024-1476
(310) 206-9103; Fax: (310) 206-6489
Email: elj@law.ucla.edu

UCLA Journal of International Law and Foreign Affairs
UCLA School of Law
405 Hilgard Ave., Box 951476
Los Angeles, CA 90024-1476
Email: jilfa@lawnet.ucla.edu

UCLA Journal of Islamic and Near Eastern Law
UCLA School of Law
405 Hilgard Ave., Box 951476
Los Angeles, CA 90024-1476
(310) 825-5294
Email: jinel@orgs.law.ucla.edu

UCLA Pacific Basin Law Journal
UCLA School of Law
405 Hilgard Ave., Box 951476
Los Angeles, CA 90024-1476
(310) 206-6174; Fax: (310) 206-6489
Email: pblj@orgs.law.ucla.edu

United States-Mexico Law Journal
University of New Mexico School of Law
1117 Stanford Dr. N.E.
Albuquerque, NM 87131-1431
(505) 277-0080; Fax: (505) 277-0751
Email: usmlj@law.unm.edu

University of Baltimore Journal of Environmental Law
University of Baltimore School of Law
Room 103, Law Center
1420 North Charles St.
Baltimore, MD 21201-5779
(410) 837-4655; Fax: (410) 837-4450

University of Miami Inter-American Law Review
University of Miami School of Law
1311 Miller Dr., P.O. Box 248087
Coral Gables, FL 33124
(305) 284-5562; Fax: (305) 284-4138
Email: ialr@law.miami.edu

University of Pennsylvania Journal of International
  Economic Law
University of Pennsylvania Law School
3400 Chestnut St.
Philadelphia, PA 19104-6204
(215) 898-6869; Fax: (215) 573-2025
Email: jiel@law.upenn.edu

Vanderbilt Journal of Transnational Law
Vanderbilt University Law School
131 21st Ave. S.
Nashville, TN 37203-1181
(615) 322-2284; Fax: (615) 322-2354

Villanova Environmental Law Journal
Villanova University School of Law
299 North Spring Mill Road
Villanova, PA 19085
(610) 519-7046; Fax: (610) 519-6472
Email: elj@law.villanova.edu

Virginia Environmental Law Journal
University of Virginia School of Law
580 Massie Road
Charlottesville, VA 22903-1789
(434) 924-3683; Fax: (434) 982-2978
Email: velj@virginia.edu

Virginia Journal of International Law
University of Virginia School of Law
580 Massie Road
Charlottesville, VA 22903-1789
(434) 924-3415; Fax: (434) 924-3237

Washington University Global Studies Law Review
Washington University School of Law
One Brookings Dr.
Campus Box 1120
St. Louis, MO 63130-4899
(314) 935-8771; Fax: (314) 935-6493

Willamette Journal of International Law and Dispute
   Resolution
Willamette University College of Law
245 Winter St. S.E., Room 111
Salem, OR 97301-3916
(503) 370-6632; Fax: (503) 370-6315
Email: intl-law-journal@willamette.edu

William and Mary Environmental Law and Policy Review
College of William and Mary
Marshall-Wythe School of Law
P.O. Box 8795
Williamsburg, VA 23187-8795
(757) 221-3279; Fax: (757) 221-3777
Email: envlaw@wm.edu

Wisconsin International Law Journal
University of Wisconsin Law School
975 Bascom Hall
Madison, WI 53706
(608) 262-3877; Fax: (608) 262-5485

The Yale Journal of International Law
Yale Law School
P. O. Box 208215
New Haven, CT 06520-8215
(203) 432-4884; Fax: (203) 436-0992

Journal, Magazine, and Newspaper Articles and Occasional Papers

Academic journals are often paginated consecutively throughout the year; that is, if the first issue of a journal in a particular volume year ends with page 200, the second issue will begin with page 201. Thus, an article cited below as *"33 Virginia Journal of International Law 351"* appears on page 351 of volume 33 of the *Virginia Journal of International Law*. It is not necessary to know which of the issues the article appeared in, because there is only one page 353 in each volume. (Page 353 was actually in the Winter 1993 issue.) For older editions of academic journals in libraries, each volume year is typically bound in a single hard-copy volume, so there should be no confusion. Journals that are not consecutively paginated require exact date information, which is provided below. The articles listed may be found in hard copy or microform in many libraries, as reprints from the journals that published them or online.

Abramovitz, Janet N. *Imperiled Waters, Impoverished Future: The Decline of Freshwater Ecosystems*. Washington, DC: Worldwatch Paper No. 128 (1996).

Baker, Betsy. **"Legal Protections for the Environment in Times of Armed Conflict."** 33 *Virginia Journal of International Law* 351 (1993).

Barberis, Julio. **"The Development of International Law of Transboundary Groundwater."** 31 *Natural Resources Journal* 167 (1991).

Booth, William A. **"Ghost City of Mixed Poisons—NATO Bombs Left Site of Petrochemical Complex a Toxic Slough."** *Washington Post,* July 21, 1999: A-15.

Caron, David. **"The Frog That Wouldn't Leap: The International Law Commission and Its Work on International Watercourses."** 3 *Colorado Journal of International Environmental Law and Policy* 269 (1992).

Charnovitz, Steve. **"Free Trade, Fair Trade, Green Trade: Defogging the Debate."** 27 *Cornell International Law Journal* 459 (1994).

Chenevert, Donald J. Jr. **"Application of the Draft Articles on the Non-navigational Uses of International Watercourses to the Water Disputes Involving the Nile River and the Jordan River."** 6 *Emory International Law Review* 495 (1992)

Costa, Pascale. **"Les effets de la guerre sur les traites relatifs au Danube, dans le cadre d'une etude globale du droit conventionnel du Danube,"** In *The Legal Regime of International Rivers and Lakes/Le regime juridique des fleuves et des lacs internationaux,* 203, 205. (Ralph Zacklin and Lucius Caflisch, eds., Martinus Nijhoff, 1981).

Dellapenna, Joseph W. **"Treaties as Instruments for Managing Internationally-Shared Water Resources: Restricted Sovereignty v. Community of Property."** 26 *Case Western Reserve Journal of International Law* 27 (1994).

————. "The Two Rivers and the Lands Between: Mesopotamia and the International Law of Transboundary Waters." 10 *Brigham Young University Journal of Public Law* 213 (1996).

Drumbl, Mark. "Waging War Against the World: The Need to Move from War Crimes to Environmental Crimes." 22 *Fordham International Law Journal* 122 (1998).

Durning, Alan Thein. *Guardians of the Land: Indigenous Peoples and the Health of the Earth.* Washington, DC: Worldwatch Paper No. 112 (1992).

Eckstein, Gabriel. "Application of International Water Law to Transboundary Groundwater Resources, and the Slovak-Hungarian Dispute over Gabčíkovo-Nagymaros." 19 *Suffolk Transnat'l Law Review* 67 (1995).

Fair, Karen. "Environmental Compliance in Contingency Operations: In Search of a Standard?" 157 *Military Law Review* 112 (1998).

Flavin, Christopher and Odil Tunali. *Climate of Hope: New Strategies for Stabilizing the Earth's Atmosphere.* Washington, DC: Worldwatch Paper No. 130 (1996).

French, Hilary F. *Green Revolutions: Environmental Reconstruction in Eastern Europe and the Soviet Union.* Washington: Worldwatch Paper No. 99 (1990).

————. *Partnership for the Planet: An Environmental Agenda for the United Nations.* Washington, DC: Worldwatch Paper No. 126 (1995).

Galambos, Judit. *Political Aspects of an Environmental Conflict: The Case of the Gabčíkovo-Nagymaros Dam System.* Perspectives of Environmental Conflict and International Relations, Jyrki Kakonen ed. (1992) 75–76.

Garibaldi, Oscar M. *The Legal Status of General Assembly Resolutions: Some Conceptual Observations.* Proceedings of the American Society of International Law 324 (1979).

Handl, Gunther. "The International Law Commission's Draft Articles on the Law of International Watercourses (General Principles and Planned Measures): Progressive or Retrogressive Development of International Law?" 3 *Colorado Journal of International Environmental Law and Policy* 132 (1992).

Hardin, Garret. "The Tragedy of the Commons." 162 *Science* 1243 (1968).

Heywood, Peter, and Karoly Ravasz. "Danube Diversion Stirs Controversy." *Engineering News Record,* February 9, 1989: 23.

Kiss, Alexandre. "The Protection of the Rhine Against Pollution." 25 *Natural Resources Journal* 613 (1985).

Liska, Miroslav B. "The Gabčíkovo-Nagymaros Project—Its Real Significance and Impacts." 6 *Europa Vincet* 7, November, 1992.

Low, Luan, and David Hodgkinson. "Compensation for Wartime Environmental Damage: Challenges to International Law After the Gulf War." 35 *Virginia Journal of International Law* 405 (1995).

Mahler, Vincent A. "The Political Economy of North-South Commodity Bargaining: The Case of the International Sugar Agreement." 38 *International Organizations* 709 (1984).

Masters, Suzette Brooks. "Environmentally Induced Migration: Beyond a Culture of Reaction." 14 *Georgetown Immigration Law Journal* 855 (2000).

McCaffrey, Stephen C. "The International Law Commission Adopts Draft Articles on International Watercourses" 89 *American Journal of International Law* 395 (1995).

———. "International Organizations and the Holistic Approach to Water Problems." 31 *Natural Resources Journal* 139 (1991).

McDougal, Myres S. *Contemporary Views on the Sources of International Law: The Effect of U.N. Resolutions on Emerging Legal Norms* (Discussion). Proceedings of the American Society of International Law 327 (1979).

Monahan, Katherine E. "U.S. Sugar Policy: Domestic and International Repercussions of Sour Law." 15 *Hastings International and Comparative Law Review* 325 (1992).

Morris, Virginia. "Protection of the Environment in Wartime: The United Nations General Assembly Considers the Need for a New Convention." 27 *International Lawyer* 775 (1993).

"Oh, Sweet Reason: A Report Counts the Cost of Europe's Sugar Subsidies on Poor Countries." *The Economist*, 73, 17 April 2004.

Okorodudu-Fubara, Margaret T. "Oil in the Persian Gulf: Legal Appraisal of an Environmental Warfare." 23 *St. Mary's Law Journal* 123 (1991).

Postel, Sandra. *Dividing the Waters: Food Security, Ecosystem Health, and the New Politics of Scarcity.* Washington, DC: Worldwatch Paper No. 132 (1996).

Riechel, Sylvia M. "Governmental Hypocrisy and the Extraterritorial Application of NEPA." 26 *Case Western Reserve Journal of International Law* 115 (1994).

Roberts, Paul. "The Sweet Hereafter: Our Craving for Sugar Starves the Everglades and Fattens Politicians." *Harper's*, 54, November 1, 1999.

Roodman, David Malin. *Paying the Piper: Subsidies, Politics, and the Environment.* Washington, DC: Worldwatch Paper No. 133 (1996).

Sachs, Aaron. *Eco Justice: Linking Human Rights and the Environment.* Washington, DC: Worldwatch Paper No. 127 (1995).

Sands, Philippe, Moderator. "The Gulf War: Environment As a Weapon." 85 *American Society of International Law Proceedings* 214 (1991).

Schmetzer, Uli. "Serbs Allege NATO Raids Caused Toxic Catastrophe: Bombed Refineries, Plants Spewed Stew of Poisons, They Say." *Chicago Tribune*, 1, July 8, 1999.

Schmitt, Michael N. "Green War: An Assessment of the Environmental Law of International Armed Conflict." 22 *Yale Journal of International Law* 1:14 (1997).

Schwabach, Aaron. "Diverting the Danube: the Gabčíkovo-Nagymaros Dispute and International Freshwater Law." 14 *Berkeley Journal of International Law* 290 (1996).

———. "Environmental Damage Resulting From the NATO Military Action Against Yugoslavia." 25 *Columbia Journal of Environmental Law* 117 (2000).

Schwebel, Stephen M. *The Effect of General Assembly Resolutions on Customary International Law.* Proceedings of the American Society of International Law 301 (1979).

Serenyi, Juliet. **"Danube Project Sours."** *Christian Science Monitor,* 9 December 1992: 19.

Sharp, Maj. Walter G., Sr. **"The Effective Deterrence of Environmental Damage During Armed Conflict: A Case Analysis of the Persian Gulf War."** 137 *Military Law Review* 1 (1992).

Smith, Ian. **"UNCTAD: Failure of the UN Sugar Conference."** 19 *Journal of World Trade Law* 296 (1985).

"The Spoils of War: What Can the Past Tell About the Effect of Military Conflict on the Environment?" *Economist* 27 March 2003.

Teclaff, Ludwik A. **"Fiat or Custom: The Checkered Development of International Water Law."** 31 *Natural Resources Journal* 45 (1991).

Teclaff, Ludwik A., and Eileen Teclaff. **"Transboundary Toxic Pollution and the Drainage Basin Concept."** 25 *Natural Resources Journal* 589 (1985).

Uhlman, Eva M. Kornicker. **"State Community Interests, Jus Cogens and Protection of the Global Environment: Developing Criteria for Peremptory Norms."** 11 *Georgetown International Law Review* 101 (1998).

Vandevelde, Kenneth. **"International Regulation of Fluorocarbons."** 2 *Harvard Environmental Review* 474 (1978).

Vargha, Janos. *Egyre tavolabb a jotol.* Valosag, November, 1981.

Wescoat, James L. Jr. **"Beyond the River Basin: The Changing Geography of International Water Problems and International Watercourse Law."** 3 *Colorado Journal of International Environmental Law and Policy* 301 (1992).

Whitaker, Richard. **"Environmental Aspects of Overseas Operations."** 1995-APR *Army Lawyer* 27, April 1995.

Williams, Paul R. **"International Environmental Dispute Resolution: The Dispute Between Slovakia and Hungary Concerning Construction of the Gabčíkovo and Nagymaros Dams."** 19 *Columbia Journal of Environmental Law* 1 (1994).

Williams, Sharon. **"Public International Law Governing Transboundary Pollution."** 13 *University of Queensland Law Journal* 112 (1984).

Xue, Hanqin. **"Relativity in International Water Law."** 3 *Colorado Journal of International Environmental Law and Policy* 45 (1992).

Treaties and Other International Agreements

Most of the treaties and other documents listed here are available on the Internet, often in multiple locations. The URLs listed after each treaty were correct as of September 10, 2004; however, Web addresses and Web content do change over time, so you may not be able to find some of these documents at the addresses given. Typing the title of the document into Google™ or a similar search engine will usually locate the document. A bound-volume set of international environmental law documents such as the *Supplement of Basic Documents to International Environmental Law and World Order* described above, compiled by Professors Guruswamy, Palmer, Weston, and Carlson, is also a useful resource.

In the citations below, U.N.T.S. stands for the United Nations Treaty Service, available in law libraries and U.N. document collections and online (a fee is charged for online access, but sample access is available free of charge). U.S.T. stands for United States Treaties and Other International Agreements, while T.I.A.S. stands for Treaties and Other International Acts Series. Both are publications containing treaties to which the United States is a party and are available in many libraries. I.L.M. stands for International Legal Materials, a series published privately by the American Society of International Law. I.L.M. is also available in libraries and from proprietary databases.

Antarctic Treaty, Dec. 1, 1959, 12 U.S.T. 794, 402 U.N.T.S. 71. Available at http://www.unog.ch/frames/disarm/distreat/antarc.htm

Cartagena Protocol on Biosafety to the Convention on Biological Diversity, Jan. 29, 2000, 39 I.L.M. 1027. Available at http://www.biodiv.org/biosafety/protocol.asp.

Charter of the United Nations, art. 2(4), June 26, 1945, 59 Stat. 1031, T.S. No. 933, 3 Bevans 1153, 1976 Y.B.U.N. 1043. Available at http://www.un.org/aboutun/charter.

Convention Concerning the Protection of the World Cultural and Natural Heritage, Nov. 16, 1972, 27 U.S.T. 37, 1037 U.N.T.S. 151. Available at http://whc.unesco.org/world_he.htm.

Convention on the Conservation of Antarctic Marine Living Resources, May 20, 1980, T.I.A.S. No. 10240, 19 I.L.M. 841 (1980). Available at http://www.ccamlr.org/pu/E/pubs/bd/pt1p1.htm.

Convention for the Amelioration of the Condition of the Wounded, Sick, and Shipwrecked Members of the Armed Forces at Sea, Aug. 12, 1949, 6 U.S.T. 3217, 75 U.N.T.S. 85 [Geneva Convention II]. Available at http://www.icrc.org/ihl.nsf/1595a804 df7efd6bc125641400640d89/44072487ec4c2131c125641e004a9977 ?OpenDocument.

Convention for the Conservation of Antarctic Seals, Feb. 11, 1972, 29 U.S.T. 441, 11 I.L.M. 251 (1972). Available at http://sedac. ciesin.org/entri/texts/antarctic.seals.1972.html.

Convention for the Prevention of Marine Pollution from Land-Based Sources, June 4, 1974, 13 I.L.M. 352 (1974). Available at http://sedac.ciesin.org/entri/texts/marine.pollution.land.based. sources.1974.html.

Convention for the Protection of the Rhine Against Chemical Pollution, Dec. 3, 1976, 16 I.L.M. 242 (1977). Available at http:// fletcher.tufts.edu/multi/texts/BH697.txt.

Convention on the Continental Shelf, June 10, 1964, 15 U.S.T. 471, 499 U.N.T.S. 311. Available at http://www.un.org/law/ilc/ texts/contsh.htm.

Convention on Long-Range Transboundary Air Pollution, Nov. 13, 1979, 1302 T.I.A.S. No. 10541, U.N.T.S. 217, 18 I.L.M. 1442 (1979). Available at http://www.unece.org/env/lrtap/full% 20text/1979.CLRTAP.e.pdf.

Convention on the Prevention of Marine Pollution by Dumping of Wastes and Other Matter, Dec. 29, 1972, 26 U.S.T. 2403, 1046 U.N.T.S. 120. Available at http://www.londonconvention.org/ documents/lc72/LC1972.pdf.

Convention on the Protection of the Rhine, Apr. 12, 1999. Available from www.iksr.org/GB/bilder/pdf/convention_on_ the_protection_of__the_rhine.pdf (visited September 27, 2004).

Convention on the Prohibition of Military or Any Other Hostile Use of Environmental Modification Techniques, Dec. 10, 1976, 31 U.S.T. 333, 1108 U.N.T.S. 151. Available at http://www.unog.ch/ frames/disarm/distreat/environ.pdf.

Convention on Wetlands of International Importance Especially as Waterfowl Habitat, Feb. 2, 1971, 996 U.N.T.S. 245, reprinted in 11 I.L.M. 963 (1972). Available at http://portal.unesco.org/en/ev. php-URL_ID=15398&URL_DO=DO_TOPIC&URL_SECTION=201.html.

Convention Relative to the Protection of Civilian Persons in Time of War, Aug. 12, 1949, 6 U.S.T. 3516, 75 U.N.T.S. 287 [Geneva Convention IV]. Available at http://www.icrc.org/ihl.nsf/0/6756482d86146898c125641e004aa3c5?OpenDocument.

Convention Relative to the Treatment of Prisoners of War, Aug. 12, 1949, 6 U.S.T. 3316, 75 U.N.T.S. 135 [Geneva Convention III]. Available at http://www.icrc.org/ihl.nsf/1595a804df7efd6bc125641400640d89/6fef854a3517b75ac125641e004a9e68?Open Document.

Espoo Convention on Environmental Impact Assessment in a Transboundary Context, Feb. 25, 1991, 30 I.L.M. 800 (1991). Available at http://www.unece.org/env/eia/documents/conventiontextenglish.pdf.

Kuwait Regional Convention for Co-operation on the Protection of the Marine Environment from Pollution, Apr. 24, 1978, 1140 U.N.T.S. 133, 17 I.L.M. 511 (1978). Available at http://sedac.ciesin.org/entri/texts/kuwait.marine.pollution.1978.html.

Montreal Protocol on Substances that Deplete the Ozone Layer, Sept. 16, 1987, 26 I.L.M. 1550 (1987). Available at http://www.unep.org/ozone/pdf/Montreal-Protocol2000.pdf

Protocol Additional to the Geneva Conventions of 12 August 1949, and Relating to the Protection of Victims of International Armed Conflicts, June 8, 1977, 1125 U.N.T.S. 3 [Protocol I]. Available at http://www.icrc.org/ihl.nsf/1595a804df7efd6bc125641400640d89/f6c8b9fee14a77fdc125641e0052b079? OpenDocument.

Protocol Additional to the Geneva Conventions of 12 August 1949, and Relating to the Protection of Victims of Non-International Armed Conflicts, June 8, 1977, 1125 U.N.T.S. 609 [Protocol II]. Available at http://www.icrc.org/ihl.nsf/1595a804df7efd6bc1256

41400640d89/d67c3971bcff1c10c125641e0052b545?Open Document.

Protocol on Environmental Protection to the Antarctic Treaty, Oct. 4, 1991, 30 I.L.M. 1461 (1991). Available at http://sedac.ciesin. org/entri/texts/antarctic.treaty.protocol.1991.html.

Rome Statute on the International Criminal Court, U.N. Doc. A/CONF. 183/9 (1998). Available at http://www.un.org/law/ icc/statute/romefra.htm.

Special Agreement for Submission to the International Court of Justice of the Differences Between Them Concerning the Gabcikovo-Nagymaros Project, Apr. 7, 1993, Hung.-Slovakia, 32 I.L.M. 1293 (1993). Available at http://www.icj-cij.org/icjwww/ idocket/ihs/ihsjudgement/ihs_ijudgment_970925.html.

Statute of the International Court of Justice, 1976 Y.B.U.N. 1052, 59 Stat. 1031, T.S. No. 993. Available at http://www.icj-cij.org/ icjwww/ibasicdocuments/Basetext/istatute.htm.

Stockholm Convention on Implementing International Action on Certain Persistent Organic Pollutants, May 22, 2001, 40 I.L.M. 532 (2001). Available at http://www.pops.int/documents/ convtext/convtext_en.pdf.

United Nations Convention on the Law of the Non-navigational Uses of International Watercourses, G.A. Res. 51/229, U.N. GAOR, 51st Sess., May 21, 1997; 36 I.L.M. 700 (1997). Available at http://www.un.org/law/ilc/texts/nonnav.htm.

United Nations Convention on the Law of the Sea, Dec. 10, 1982, U.N. Doc. A/CONF.62/122, 21 I.L.M. 1261 (1982). Available at http://www.un.org/Depts/los/convention_agreements/texts/ unclos/unclos_e.pdf.

United Nations Economic Commission for Europe Convention on the Protection and Use of Transboundary Watercourses and International Lakes, Mar. 17, 1992, 31 I.L.M. 1312. Available at http://www.unece.org/env/water/pdf/watercon.pdf.

United Nations Framework Convention on Climate Change, May 9, 1992, 31 I.L.M. 849 (1992). Available at http://unfccc.int/ resource/docs/convkp/conveng.pdf.

Other International Materials

The documents listed here are of various types and are not always easy to locate; to aid in the search, web addresses have been provided for nearly all. All decisions of the International Court of Justice, along with many other ICJ materials, are available from the ICJ website at www.icj-cij.org.

Affaire du Lac Lanoux (Spain v. Fr.), 12 Reports of International Arbitral Awards 281 (1957), digested in 53 Am. Journal of International Law 156 (1959).

Corfu Channel Case (U.K. v. Alb.), 1949 I.C.J. 4, 22 (1949) (determination on the merits). Available at http://www.icj-cij.org/icjwww/icases/icc/icc_ijudgment/iCC_ijudgment_19490409.pdf.

Final Report to the Prosecutor by the Committee Established to Review the NATO Bombing Campaign Against the Federal Republic of Yugoslavia. Available at http://www.un.org/icty/pressreal/nato 061300.htm.

Gabčíkovo-Nagymaros Dispute (Slovakia v. Hungary), 1997 I.C.J. 7 (1997). Available at http://www.icj-cij.org/icjwww/idocket/ihs/ihsjudgement/ihs_ijudgment_970925_frame.htm

Johannesburg Plan of Implementation, Sep. 4, 2002, U.N. Doc. A/CONF.199/20. Avaliable at http://www.un.org/esa/sustdev/documents/WSSD_POI_PD/English/POIToc.htm

Legality of the Threat or Use of Nuclear Weapons (Advisory Opinion), 35 I.L.M. 809, 824 (July 8, 1996). Available at http://www.icj-cij.org/icjwww/icases/iunan/iunan_judgment_advisory%20opinion_19960708/iunan_ijudgment_19960708_Advisory%20Opinion.htm

Legality of Use of Force (Yugo. v. Belg.), 1999 I.C.J. (Apr. 29, 1999), (Application of Yugoslavia). Available at http://www.icj-cij.org/icjwww/idocket/iybe/iybeapplication/iybe_application_19990428.html (visited 10 September 2004). This is the document initiating the first (in alphabetical order) of the ten lawsuits against NATO members; each of the ten applications and requests for provisional measures is identical, *mutatis mutandis.* The other countries sued by Yugoslavia were Canada, France, Germany, Italy, the Netherlands, Portugal, Spain, the United Kingdom, and the United States; the documents relating to these suits are also available at http://www.icj-cij.org/icjwww/idocket.htm.)

Making Sustainable Commitments: An Environment Strategy for the World Bank. Available at http://www-wds.worldbank.org/servlet/WDS_IBank_Servlet?pcont=details&eid=000094946_01110704111523.

Non-Binding Authoritative Statement of Principles for a Global Consensus on the Management, Conservation and Sustainable Development of all Types of Forests, (Aug. 14, 1992), U.N. Doc. A/CONF.151/26 (Vol. III). Available at http://www.un.org/documents/ga/conf151/aconf15126–3annex3.htm.

Permanent Court of Arbitration. *Optional Rules for Arbitration of Disputes Relating to Natural Resources and/or the Environment* (June 19, 2001). Available at http://www.pca-cpa.org/ENGLISH/EDR/

Report of the United Nations Stockholm Conference on the Human Environment, U.N. Doc. A/CONF.48/14/Rev. 1 (1973), 11 I.L.M. 1416 (1972). Available at http://www.unep.org/Documents/Default.asp?DocumentID=97.

Rio Declaration on Environment and Development, (June 14, 1992), U.N. Doc. A/CONF.151/26 (vol. I), 31 I.L.M. 874 (1992). Available at http://www.un.org/documents/ga/conf151/aconf15126–1annex1.htm.

S.C. Res. 687, U.N. SCOR, 2981st mtg., U.N. Doc. S/RES/687 (1991). Available at http://www.un.org/Docs/sc/committees/IraqKuwait/IraqResolutionsEng.htm

Trail Smelter Case (U.S. v. Can.), 3 R.I.A.A. 1905, 1965 (1941), reprinted in 35 American Journal of International Law 684 (1941).

UNEP, *The Kosovo Conflict: Consequences for the Environment and Human Settlements* (1999). Available at http://www.grid.unep.ch/btf/final/index.html.

United Nations Conference on Environment and Development: Agenda 21, (June 13, 1992), U.N. Doc. A/CONF.151/26 (1992). Available at http://www.un.org/esa/sustdev/documents/agenda21/english/agenda21toc.htm.

World Charter for Nature, (Oct. 28, 1982), G.A. Res. 37/7 (Annex), U.N. GAOR, 37th Sess., Supp. No. 51, at 17, U.N. Doc. A/37/7, 22 I.L.M. 455 (1983). Available at http://www.un.org/documents/ga/res/37/a37r007.htm.

U.S. Materials

The U.S. legal system is complex; in addition to Congress and fifty state legislatures enacting statutes and the president and fifty state governors issuing various proclamations, orders, and decrees, there are several hundred state and federal courts issuing reported decisions that become part of our common law. Lawyers spend years in law school learning how all of these pieces fit together.

The statutes, judicial opinions, and administrative materials listed below may be repealed, overturned, superseded, or otherwise rendered obsolete at any time. Citator services such as Shepard's Citations and West's Keycite keep track of these changes, but correctly interpreting the information provided is difficult; it is probably best to ask a law librarian for assistance.

Reported court decisions are bound in volumes called reporters; a full set of case reporters should be available in any law library and in many other libraries. In the citations below, So. 2d stands for *Southern Reporter, Second Series*; this is a reporter of cases from state (but not federal) courts in several southern states. O.R. stands for *Ontario Reports*. F. Supp. stands for *Federal Supplement*, a reporter of cases from federal trial (district) courts. F.2d and F.3d stand for *Federal Reporter, Second Series* and *Federal Reporter, Third Series*, respectively; they contain decisions of federal appellate (circuit) courts.

Fla. Dist. Ct. App. stands for Florida District Court of Appeals. (In Florida district courts are appellate courts, and circuit courts are trial courts, in a reversal of the federal courts' nomenclature.) In the federal case citations, Cir. stands for Circuit and D. stands for District; E.D. stands for Eastern District and S.D. for Southern District. Cert. stands for certiorari, one of the procedural mechanisms by which cases may come before the Supreme Court. For the statutes, Pub. L. stands for Public Law; Stat. stands for Statutes at Large; and U.S.C. stands for United States Code.

All of this may seem complicated, but any lawyer or law librarian looking at the citation will be familiar with these abbreviations and will be able to find the case or statute instantly.

Cases

Florida Sugar Cane League, Inc. v. South Florida Water Management District, 617 So. 2d 1065 (Fla. Dist. Ct. App. 1993).

Heartland By-Products, Inc. v. United States, 74 F. Supp. 2d 1324, 1326–27 (Court of International Trade 1999), http://www.cit.us-courts.gov/slip_op/Slip_op99/99–110.pdf.

Miccosukee Tribe of Indians v. Florida Department of Environmental Protection, 656 So. 2d 505 (Fla. Dist. Ct. App. 1995).

Miccosukee Tribe of Indians of Florida v. United States Environmental Protection Agency, 105 F.3d 599 (11th Cir. 1997), http://www.ca11.uscourts.gov/opinions/ops/19955080.OPA.pdf.

United States v. Ivey, 747 F.Supp. 1235 (E.D. Mich. 1990); related Canadian decision, *United States v. Ivey* [1995], 26 O.R. 3d 533 (Ont. Gen. Div.), affirmed [1996], 30 O.R. 3d 370 (Ont. Ct. App.).

United States Cane Sugar Refiners' Association v. Block, 544 F. Supp 883 (Court of International Trade 1982), affirmed, 683 F.2d 399 (Court of Customs and Patent Appeals 1982).

United States v. South Florida Water Management District, 922 F.2d 704 (11th Cir. 1991); cert. denied, 502 U.S. 953 (1991); on remand, 847 F. Supp. 1567 (S.D. Fla. 1992); affirmed in part, reversed in part, 28 F.3d 1563 (11th Cir. 1994); cert. denied 514 U.S. 1107 (1995).

Statutes and Other Materials

Environmental Effects Abroad of Major Federal Actions, Executive Order 12,114, 44 Fed. Reg. 1957 (1979), 42 U.S.C. § 4321 (1982), http://frwebgate4.access.gpo.gov/cgi-bin/waisgate.cgi?WAISdocID=88002119798+10+0+0&WAISaction=retrieve.

Foreign Sovereign Immunities Act of 1976, Pub. L. 94–583, 90 Stat. 2891, 28 U.S.C. §§ 1330, 1332(a), 1391(f) & 1601–11.

International and Operational Law Department, The Judge Advocate General's School, U.S. Army, JA 442, *Operational Law Handbook,* 18–1, 18–2 (1996).

National Environmental Policy Act, 42 U.S.C. § 4321–4375.

Policy Review Directive/NSC-23, United States Policy on Extraterritorial Application of the National Environmental Policy Act (NEPA) Apr. 8, 1993.

Web Resources

The websites listed here are commercial database providers; they store a wide variety of legal materials. In addition to these sites, the websites of the organizations listed in Chapter 7 are a valuable research tool; the various United Nations websites, in particular, contain almost all of the treaties, international cases, and other documents that the researcher in international environmental law is likely to need.

ENTRI—Environmental Treaties and Resource Indicators: http://sedac.ciesin.columbia.edu/entri/index.jsp

ENTRI provides the text and current status of almost all multilateral environmental treaties, plus a search engine.

Environmental Policy Index (formerly Environmental Knowledgebase Online): www.ebsco.com/home/

The Environmental Policy Index offers online access to a wide variety of scientific and other scholarly papers. A fee is charged for access.

Findlaw: www.findlaw.com

Findlaw offers articles and summaries on a wide variety of legal topics. There is relatively little coverage so far of international environmental law, but new material is constantly being added.

Hein Online: heinonline.org

Hein Online offers access to the full texts of more than 500 legal journals, as well as treaties and federal administrative materials. A fee is charged for access.

Lexis: www.lexis.com

Lexis is a provider of legal research databases to lawyers, law students, scholars, and judges. It includes several valuable international environmental resources, including treaties, cases, foreign law databases (including extensive coverage of the laws of English-speaking countries and of the European Union), and *Interna-*

tional Legal Materials, a publication of the American Society of International Law containing treaties, cases, and other useful documents. The coverage of Lexis is similar, but not identical, to that of its competitor Westlaw; Westlaw has more extensive coverage in some areas, while Lexis has more extensive coverage in others. A fee is charged for access.

United Nations Treaty Collection: untreaty.un.org

The United Nations Treaty Collection includes the text and ratification status of treaties deposited with the secretary-general of the United Nations in accordance with Section 102 of the U.N. Charter, which provides that every treaty and every international agreement entered into by any member of the United Nations after the present charter comes into force shall as soon as possible be registered with the secretariat and published by it. A fee is charged for access, but limited sample access is available.

Westlaw: www.westlaw.com

Westlaw, like Lexis, is a provider of legal research databases to lawyers, law students, scholars, and judges. It includes several valuable international environmental resources, including, among others, databases for decisions of the International Court of Justice, treaties to which the United States is a party, materials of several foreign countries, journal articles, the *Restatement of Foreign Relations,* and *International Legal Materials.* The coverage of Westlaw is similar but not identical to that of its competitor Lexis. A fee is charged for access.

Index

Superfund Act (CERCLA),
114–115
Sustainable development, 33–37, 39
Commission on Sustainable
Development (CSD), 25, 150
definition of (UN report), 163
Gabcíkovo-Nagymaros case and,
152, 215–220
intergovernmental
organizations for, 259–260
Rio Declaration, as goal of, 62
World Bank manifesto and, 155
World Summit on (2002), 23,
155
Swaziland, subsidized sugar and,
64–66
Szigetköz, 52, 53

Taking of land, 108
Temperature (global warming),
73–79
Territorial sovereignty, 16, 17, 33
Texaco Overseas Petroleum
Company (TOPCO),
arbitration, 7–8
Three-prong test, 112, 116
Tisza cyanide spill, 60–61, 87, 110,
154
Tonga, geostationary orbits and,
93
TOPCO. *See* Texaco Overseas
Petroleum Company
Toronto Atmosphere Conference,
148
Torture Victim Protection Act, 118
Tourism, as threat to ecosystems,
80–81. *See also* Ecotourism
Toxic waste
disposal on land, 86–88
from Pancevo chemical
complex destruction,
122–125

Tisza cyanide spill, 60–61, 87,
110
transboundary
shipment/disposal of, 86–88
treaties/agreements on, 86
See also Basel Convention
Trade
barriers, 64, 66–68, 118–121
embargo, 115
free, 115, 119
General Agreement on Tariffs
and Trade (GATT), 115, 119
international, 62–68, 119, 125
North America Free Trade
Agreement (NAFTA), 121,
151, 258
See also World Trade
Organization (WTO)
Tragedy of the commons, 34
Trail Smelter arbitration, 14–15, 33,
39, 220–221
Transborder environmental harm,
14–15, 19, 21–22
Transboundary environmental
resources, 16–17, 32
Transboundary pollution, 32,
70–79
Transboundary
shipment/disposal of toxic
waste, 86–88
Transboundary watercourses. *See*
Watercourses, international
Treaties, 23–24, 190–214, 308–311
Antarctic Treaty, 140, 190–191
Antarctic Treaty Environmental
Protection Protocol, 192–193
Basel and Bamako
Conventions, 193–200
biodiversity, 25
Boundary Waters Treaty,
136–137
climate change, 25

About the Author

Aaron Schwabach is a professor of law at Thomas Jefferson School of Law in San Diego. He is the author of numerous scholarly articles and other published works on international environmental law. Dr. Schwabach has taught classes in the subject since 1991.